ostp/big_data_fact_sheet_final_1.pdf). Australia published "The Australian Public Service Big Data Strategy" in 2013. Subsequently, the Chinese government began emphasizing big data as one of the strategic resources of social development in 2013 and issued the "The Action Plan for Promoting Big Data Development" in 2015, including a proposal for "Developing Science Big Data". In 2012, the UN Global Pulse published "Big Data for Development: Opportunities and Challenges" to promote the significant role of big data in responding to climate changes. The International Council for Science (joined in 2017 with the International Social Science Council to form the International Sciences Council) published their "Strategic Plan 2012–2017", which emphasized the importance of data management in new knowledge discovery.

Big data has created a new computational perspective in the use of continuously collected data from various sources to explore trends in large volumes of data and to better understand world dynamics. Such advances bring great opportunities for Digital Earth to play its visionary role in integrating the massive amount of multidimensional, multitemporal, and multiresolution geospatial data as well as socioeconomic data in a framework for comprehensive analysis and application systems about the Earth.

Digital Earth has evolved into a new connotation of 'big Earth data'. Big Earth data incorporates the litany of powerful tools requisite to understanding and explaining the Earth system and to investigating sustainable global development. It focuses on the synthesis and systematic observation of Earth, as well as data-intensive methods for studying Earth system models with the goal of increasing knowledge discovery. Big Earth data can be expected to promote the Digital Earth vision by connecting multiple satellites and geographical information centers that rely on national spatial infrastructures and high-speed internet to complete the acquisition, transmission, storage, processing, analysis, and distribution of spatial data.

1.5 Relationship with Other Initiatives

1.5.1 Geospatial Information Infrastructures

The United States pioneered the development and implementation of the National Spatial Data Infrastructure (NSDI) in the 1990s. Clearly defined, the NSDI is the sum of the technologies, policies, standards, human resources, and related activities necessary to collect, process, publish, use, maintain, and manage geospatial data from all levels of government, private and nonprofit organizations, and academic communities. The NSDI makes existing and accurate geospatial data more accessible, greatly facilitates the collection, sharing, distribution and utilization of geospatial data, and has played an active role in economic growth, environmental quality and protection, and social progress in the United States.

The NSDI model has been accepted and adopted to fit the needs of many other countries that have implemented their own spatial data infrastructure plans. The federal government of Canada implemented the GeoConnections program in 1999, a national program of partnerships involving the federal government, provincial (district) governments, municipal and local governments, research institutes, universities, and private companies. The main role of GeoConnections is to establish the Canadian Geospatial Data Infrastructure (CGDI) and enable online access to Canadian geospatial databases and services. The CGDI is the sum of the policies, technologies, standards, access systems and protocols necessary to coordinate all geospatial databases in Canada and make them available on the internet. For more than a decade, the implementation of GeoConnections has enabled online access to Canadian geospatial databases and services, and effectively coordinated partnerships, investments and developments between federal, provincial, local government, private and academic communities.

In 2007, the Infrastructure for Spatial Information in Europe (INSPIRE) Directive came into force (https://inspire.ec.europa.eu/about-inspire/563). This directive established a web-based infrastructure to make more visible, shareable and usable environmental and geospatial information necessary to support European environmental policies that affect the environment such as transport, agriculture, and marine policy. INSPIRE is decentralized, i.e., the infrastructure builds on those set up and maintained by the 28 EU member states. It does not require the collection of new data and develops the technical, and organizational arrangements to achieve interoperability among the infrastructures in the member states and among the 34 data themes falling in the scope of the directive. INSPIRE will take more than 12 years to implement, from 2007 when the directive was adopted to 2019–20 and beyond. As this process takes place, it is important to consider the technological and policy developments that will shape the future data infrastructures so that the investments of today are open to the developments of tomorrow.

1.5.2 Earth Observation Program

Earth observation has been become a major part of many countries' environmental and defense programs since the final decades of the last century. Nations were influenced by the Planetary Mission of NASA's Earth Observation Program. The program was developed for the scientific research of the Earth systems. Its goals were to collect sufficient data on the Earth's systems to enable whole planetary assessments and conduct comprehensive research on the Earth. NASA's program consists of three parts: the scientific plan, Earth observation platforms, and data information systems.

A new generation of space-Earth observation continues and has been extended to incorporate observations of the land, atmosphere, ocean, ecosystem processes, water and energy cycles, and solid Earth.

The Global Monitoring for Environment and Security (GMES) program was jointly established by the European Space Agency (ESA) and the European Commission in 2003. The ESA created a series of next-generation Earth observation missions, including the Copernicus program. To meet the operational needs of the Copernicus program, the ESA developed the Sentinel program to replace older Earth observation missions. Each Sentinel mission is based on a paired satellite model to provide datasets for Copernicus Services and focuses on different aspects of Earth observation, including atmospheric, oceanic, and land monitoring.

Earth observations have expanded rapidly around the globe, as demonstrated by the fact that the GEO now has more than 100 member nations. Bringing Earth observation down to Earth with an ever-increasing number of Earth observation satellites with increasing spatial, temporal and spectral resolutions represents a critical data input to the Digital Earth concept.

1.5.3 National/Regional Digital Earth Programs

Dozens of countries such as Australia, China, Japan, Singapore, South Africa, and the European Commission have generated their own Digital Earth-related programs. There has been important progress in these efforts, such as Digital Earth Australia established by the Australia federal government in 2017, the Geoscience Australia Data Cube (supported by the Commonwealth Scientific and Industrial Research Organization, the National Computational Infrastructure, and the National Collaborative Research Infrastructure Strategy of Australia), Digital China promoted by the Chinese government, the Key Laboratory of Digital Earth Sciences established by the Chinese Academy of Sciences, and the IDEAS (International Digital Earth Applied Science Research Center) at Chubu University in Japan, as well as those at several universities with Digital Earth departments or laboratories (e.g., Austria and Malaysia). Some of these are described in detail in Part III of this manual.

1.6 Digital Earth in Response to Global Challenges

Correlated with and a derivative of many sciences dealing with the surface and near-surface of the Earth, Digital Earth was envisioned as an initiative for harnessing the Earth's data and information resources. With powerful tools to quantitatively describe a science-based representation of the planet, Digital Earth could serve as a tool to map, monitor, measure, and forecast natural and human activities. The prowess of the Digital Earth technology was envisioned as requisite to assist nations, organizations, and individual citizens in addressing the problems humans are facing in the 21st century. These challenges for all nations, such as climate change, natural disasters, and sustainable development, require the comprehensive scope and analytical capacity of Digital Earth technology.

1.6.1 Response to Climate Change

Since the middle of the 20th century, large-scale, high-intensity human activities and the rapid growth of the population and social economy have compounded global change problems such as global warming, air pollution, water pollution, land degradation, rapacious resource exploitation, and biodiversity decline. Global change threatens national security and all aspects of our lives, including economic and social development conditions from social, economic, living, and health perspectives. Sustainable development is now recognized as the most serious challenge facing human society.

The United Nations, in partnership with various intergovernmental coalitions, has organized and implemented a series of environmental research programs on global or regional scales, such as the World Climate Programme (WCP), the Man and the Biosphere Programme (MAB), and the International Biosphere Programme (IBP). Within the International Geosphere Biosphere Programme (IGBP), each country is challenged to address the natural resource and environmental issues caused by global change as a primary means to achieve sustainable approaches for socioeconomic development.

Global change is recognized as a significant threat to sustainable development worldwide. To address these multidisciplinary issues at a global scale, global change research faces the unpredicted challenge of obtaining copious data from the interacting subsystems of the Earth for analytical modeling and generating management decisions (Chen 1999; Shupeng and van Genderen 2008). Thus, it is important that Digital Earth facilitates the collection of data from various elements of the Earth system through monitoring the progress of global change in large-scale, long-term sequences, and aids in data processing, analysis, and simulation.

The Paris Agreement, which was negotiated by representatives of 196 countries, was endorsed at the 21st Conference of the Parties of the United Nations Framework Convention on Climate Change (UNFCCC) in LeBourget, France on 12 December 2015. The Paris Agreement's long-term goal is to "strengthen the global response to the threat of climate change by keeping a global temperature rise this century well below 2 degrees Celsius above preindustrial levels and to pursue efforts to limit the temperature increase even further to 1.5 degrees Celsius." The agreement states the need to "strengthen the ability of countries to deal with the impacts of climate change" as well as reduce the risks and effects of climate change (https://unfccc.int/process-and-meetings/the-paris-agreement/the-paris-agreement). Under this agreement, starting in the year 2020, a litany of financial policies and new technology frameworks will be put into action to support the realization of greenhouse gas emission mitigation, adaptation, and finance. Increasingly, scientists are documenting that the impacts of temperature increases in the polar regions indicate that our collective actions may be too little, too late.

Big Earth data should provide a wide range of long-term sequences and multiple spatiotemporal scales to cover all of Earth's systems including the atmosphere, cryosphere, hydrosphere, lithosphere, and biosphere. To stock Digital Earth with big

Earth data requires the only known science approach, which is to amass a space-air-ground integrated Earth observation system and a global near-real time, all-weather Earth data acquisition network. Through continuous and long-term monitoring of the Earth system, scientists can use advanced geospatial processing technologies to simulate and analyze Earth's dynamic surface processes and reveal spatiotemporal change mechanisms. Stakeholders will need to formulate scientific strategies and take progressive actions to respond to global change for sustainable development at varying local and regional scales. In this sense, big Earth data provides strong support to the Digital Earth vision, which will hopefully strengthen new approaches to global change research.

1.6.2 Response to the UN SDGs

In September 2015, the United Nations General Assembly adopted "Transforming our World: The 2030 Agenda for Sustainable Development" (United Nations 2015). This international milestone provides a blueprint for action for all countries and stakeholders. This agenda defines 17 Sustainable Development Goals (SDGs) and 169 targets and creates a global indicator framework until 2030. The 2030 Agenda for Sustainable Development provides a new insight into the global actions and transformative policies necessary to guide our collective pursuit of sustainable development.

Achieving sustainable development presents all countries with a set of significant development challenges. These challenges are inherently embedded with spatial-temporal complexities, that is, they are almost entirely geographic in nature. Many of the structural issues impacting sustainable development goals can be analyzed, modeled, and mapped using Earth observation data, which can provide the integrative and quantitative framework necessary for global collaboration, consensus and evidence-based decision making.

Digital Earth is closely interrelated with the global sustainable development challenges and processes, as evidenced through national Earth observation agencies' efforts to connect and integrate big Earth data into the application of many social and environmental programs. Earth observation data provide a substantial contribution to the achievements of the SDGs in support of decision making by monitoring impacts and results, improving the standardization of national statistics, addressing cross-cutting themes such as climate and energy, and facilitating countries' approaches to working across different development sectors (Anderson et al. 2017).

At the United Nations World Geospatial Information Congress (UNWGIC) held in Deqing, Zhejiang Province in China from 19 to 21 November 2018, attention was paid to strengthening national geospatial information management and systems and national implementation of the 2030 Agenda for Sustainable Development (https://www.unwgic2018.org/). It has become important to the science and governance communities to understand, analyze and discover knowledge from huge geospatial data resources. This must be accomplished in a collaborative way among nations to effectively address local, regional, and global challenges and to share big Earth data worldwide as prerequisites to meet the requirements of sustainable development.

Effective transfer of all relevant technologies and Earth-related data represents an important challenge (Scott and Rajabifard 2017). However, under the Digital Earth framework, there are immense opportunities for digital transformation and sharing of resources. Achieving sustainable development will entail significant advances in overcoming political and technical bottlenecks to smooth the digital divide. Internet-based infrastructure with advancing 5G communication shows promise for expanding Digital Earth technologies to all nations.

1.6.3 Response to Disaster Mitigation

Addressing natural and human-caused disasters remains the highest priority of all nations. Climate change experts are in agreement that global warming will increase the frequency and intensity of storms and disruptive weather patterns. Therefore, application of the Digital Earth framework and technology for disaster response and mitigation is of paramount importance.

Recently, the Sendai Framework for Disaster Risk Reduction 2015–2030 (Sendai Framework) was adopted by the UN member states in March 2015 at the World Conference on Disaster Risk Reduction held in Sendai, Japan, and endorsed by the full UN General Assembly in June 2015. This 15-year development framework agenda contains seven targets and four priorities for action. The United Nations International Strategy for Disaster Reduction (UNISDR) has been tasked with supporting the implementation, follow-up and review of the Sendai Framework. The framework's central aim is to "reduce disaster risk… and losses in lives, livelihoods and health and in the economic, physical, social, cultural and environmental assets of persons, businesses, communities and countries" with the efforts of local governments, the private sector and other stakeholders within a voluntary, nonbinding agreement (https://www.unisdr.org/we/coordinate/sendai-framework).

Notably, disaster-related applications have been prominent since the inception of the Digital Earth community. Chen's (2004) comprehensive review of Digital Earth science in China includes examples of research on flood, coastal, river, and other disasters. The International Society for Digital Earth has sponsored or cosponsored many disaster-oriented workshops and symposia. Importantly, the collaboration with UNISDR, GEO and CODATA and other international associations has been anchored by the common commitment to collaboration and focus on applications for disaster response. Chapter 15 addresses these applications.

1.7 Conclusions/Structure of the Manual

In this manual, the Digital Earth vision has been introduced in the first chapter. Part I has eleven chapters about various Digital Earth technologies; Part II has seven chapters describing the role of Digital Earth in multidomain applications; and Part III contains four chapters showing how the Digital Earth concept has developed from Al Gore's original vision to its current implementation as employed around the world through four regional/national chapters of the International Society for Digital Earth. Part IV considers Digital Earth education and ethics. The concluding chapter of this Manual of Digital Earth describes some of the key challenges and future trends for the development of Digital Earth over the coming years.

Digital Earth is an evolving concept that is strongly influenced by the evolution of technology and the availability of new data. In a couple of years, the Earth will be revisited several times a day by the new generation of satellites, and real-time observation will no longer be a chimera. As we look to the future, it is unlikely that a unified vision of Digital Earth will capture all the perspectives of all stakeholders. A one-size-fits-all Digital Earth would not be appropriate for all nations and cultures. The current social and technological trends expressed in the literature prescribe a robust and comprehensive list of likely characteristics for an updated version for Digital Earth, which closely follows the original vision. There will be a series of connected perspectives of Digital Earth based on varying priorities and applications of the same framework data sources operating with different user-specified functionalities. In the future, the concept and vision of Digital Earth will evolve with the development of science and technology.

References

Anderson K, Ryan B, Sonntag W et al (2017) Earth observation in service of the 2030 agenda for sustainable development. Geo-Spatial Inf Sci 20(2):77–96

Annoni A, Craglia M, Ehlers M et al (2011) A European perspective on Digital Earth. Int J Digit Earth 4(4):271–284

Chen S (2004) Geo-information science and Digital Earth. Science Press and Science Press USA Inc, Beijing

Chen S (1999) The 'Digital Earth' as a global strategy and its master point. J Remote Sens 3(4):247–253

Craglia M, de Bie K, Jackson D et al (2012) Digital Earth 2020: towards the vision for the next decade. Int J Digit Earth 5(1):4–21

de Longueville B, Annoni A, Schade S et al (2010) Digital Earth's nervous system for crisis events: real-time sensor web enablement of volunteered geographic information. Int J Digit Earth 3(3):242–259

Foresman TW (2008) Evolution and implementation of the Digital Earth vision, technology and society. Int J Digit Earth 1(1):4–16

Foresman TW (2010) Sustainability and an Earth Operating System for Gaia. In: Crist E, Rinker BH (eds) Gaia in turmoil. MIT Press, Cambridge, MA, pp 243–253

Giles J (2006) Wikipedia rival calls in the experts. Nature 443(7111):493

Goodchild MF (1999) Implementing Digital Earth: a research agenda. In: Xu G, Chen Y (eds) Towards Digital Earth: proceedings of the International Symposium on Digital Earth. Science Press, Beijing, pp 21–26

Goodchild MF (2000) Discrete global grids for Digital Earth. In: Proceedings of 1st international conference on discrete global grids, Citeseer, California, Santa Barbara, pp 26–28

Goodchild MF (2007) Citizens as sensors: the world of volunteered geography. GeoJournal 69(4):211–221

Goodchild MF (2008) The use cases of Digital Earth. Int J Digit Earth 1(1):31–42

Goodchild MF (2013) The quality of big (geo)data. Dialogues Hum Geogr 3(3):280–284

Goodchild MF, Guo H, Annoni A et al (2012) Next-generation Digital Earth. Proc Natl Acad Sci U S A 109(28):11088–11094

Gore A (1992) Earth in the balance: ecology and the human spirit. Houghton Mifflin, Boston

Gore A (1999) The Digital Earth: understanding our planet in the 21st Century. Photogramm Eng Remote Sens 65(5):528–530

Grossner KE, Clarke KC (2007) Is Google Earth, "Digital Earth?" – defining a vision. https://www.researchgate.net/publication/254735434_Is_Google_Earth_Digital_Earth-Defining_a_Vision. Accessed 10 Mar 2019

Grossner KE, Goodchild MF, Clarke KC (2008) Defining a Digital Earth system. Trans GIS 12(1):145–160

Guo H (2012) China's earth observing satellites for building a Digital Earth. Int J Digit Earth 5(3):185–188

Guo H, Fan X, Wang C (2009) A Digital Earth Prototype System: DEPS/CAS. Int J Digit Earth 2(1):3–15

Guo H, Liu Z, Jiang H et al (2017) Big earth data: a new challenge and opportunity for Digital Earth's development. Int J Digit Earth 10(1):1–12

Guo H, Wang L, Chen F et al (2014) Science Big Data and Digital Earth. Chin Sci Bull 59(12):1047–1054

Guo H (2018) A project on big earth data science engineering. Bull Chin Acad Sci 33(8):818–824

He G, Wang L, Ma Y et al (2015) Processing of earth observation big data: challenges and countermeasures. Chin Sci Bull 60(5–6):470–478

He G, Zhang Z, Jiao W et al (2018a) Generation of ready to use (RTU) products over China based on Landsat series data. Big Earth Data 2(1):56–64

He G, Wang G, Long T et al (2018b) Opening and sharing of big earth observation data: challenges and countermeasures. Bull Chin Acad Sci 33(8):783–790

Heipke C (2010) Crowdsourcing geospatial data. ISPRS J Photogramm Remote Sens 65(6):550–557

Howe J (2006) The rise of crowdsourcing. Wired Mag 14(6):1–4

International Society for Digital Earth (1999) Beijing Declaration on Digital Earth. Approved at the First International Symposium on Digital Earth. http://www.digitalearth-isde.org/society/56. Accessed 10 Mar 2019

International Society for Digital Earth (2009) Beijing Declaration on Digital Earth. Approved at the Six International Symposium on Digital Earth. http://www.digitalearth-isde.org/society/56. Accessed 10 Mar 2019

International Society for Digital Earth (2012) Statutes of the International Society for Digital Earth. http://www.digitalearth-isde.org/statues. Accessed 10 Mar 2019

Keysers JH (2015) Digital globe review 2015, published by the Australia and New Zealand cooperative research centre for spatial information. http://www.crcsi.com.au/assets/Uploads/Globe-review-paper-March-2015.pdf. Accessed 10 Mar 2019

Lewis A, Oliver S, Lymburner L et al (2017) The Australian geoscience data cube—foundations and lessons learned. Remote Sens Environ 202:276–292

Lu F, Zhang H (2014) Big data and generalized GIS. Geomat Inf Sci Wuhan Univ 39(6):645–654

Mahdavi-Amiri A, Alderson T, Samavati F (2015) A survey of Digital Earth. Comput Gr 53:95–117

Mueller N, Lewis A, Roberts D et al (2016) Water observations from space: mapping surface water

from 25 years of Landsat imagery across Australia. Remote Sens Environ 174:341–352
NASA (2009) New vision for 'eyes on the earth'. http://www.nasa.gov/topics/earth/features/Eyes_1.5.html. Accessed 10 Mar 2019
Sagar S, Roberts D, Bala B et al (2017) Extracting the intertidal extent and topography of the Australian coastline from a 28 year time series of Landsat observations. Remote Sens Environ 195:153–169
Sahr K, White D, Kimerling AJ (2003) Geodesic discrete global grid systems. Cartogr Geogr Inf Sci 30(2):121–134
Shupeng C, van Genderen J (2008) Digital Earth in support of global change research. Int J Digit Earth 1(1):43–65
Scott G, Rajabifard A (2017) Sustainable development and geospatial information: a strategic framework for integrating a global policy agenda into national geospatial capabilities. Geo-spatial Inf Sci 20(2):59–76
United Nations (2015) Transforming our World: The 2030 Agenda for Sustainable Development. Sustainable Development Goals Knowledge Platform. https://sustainabledevelopment.un.org/post2015/transformingourworld Accessed 18 Mar 2019
Yokokawa M (2002) Earth simulator system. http://www.thocp.net/hardware/nec_ess.html. Accessed 6 Mar 2019

Zhen Liu received her Ph.D. degree in Cartography and Geographical Information System from the Institute of Remote Sensing and Digital Earth, Chinese Academy of Sciences. She is now the Executive Director of the secretariat of the International Society for Digital Earth and the Managing Editor of the *International Journal of Digital Earth.*

John van Genderen is a Professor from the ITC, who has been involved in Digital Earth since the first ISDE Symposium in Beijing in 1999. His main research is in Remote Sensing Image Fusion, and in transfer of technology to Less Developed countries. He has carried out EO projects in some 140 countries.

Lizhe Wang is a "ChuTian" Chair Professor at School of Computer Science, China University of Geosciences (CUG). Prof. Wang received B.E. & M.E. from Tsinghua University and Doctor of Eng. from University Karlsruhe (Magna Cum Laude), Germany. His main research interests include remote sensing image processing and spatial data processing.

Part I
Technologies of DE

2

Discrete Global Grid Systems and Digital Earth

Troy Alderson, Matthew Purss, Xiaoping Du, Ali Mahdavi-Amiri and Faramarz Samavati

Abstract In this chapter, we provide a thorough discussion on Digital Earth with particular focus on Discrete Global Grid Systems (DGGS), which are a standardized representation of the Earth. We describe the necessary components of a DGGS, such as the underlying 2D representation, indexing system, projection, and cell types. We also discuss a selection of well-known public and commercial DGGSs followed by current DGGS standards.

Keywords Discrete Global Grid · OGC Standard · Digital Earth

2.1 Introduction

Digital Earth is a framework for geospatial data management. In this model, data are assigned to locations on a 3D model of the Earth and analyzed at multiple resolutions, each representing data at a specific level of detail. To locate and retrieve data sets associated with the Earth, mechanisms are needed for data representation, region addressing, and the assignment and retrieval of data for a region of interest. Digital Earth provides a reference model that can handle all of these queries.

T. Alderson · F. Samavati
Department of Computer Science, University of Calgary, Calgary, Canada
e-mail: tfalders@ucalgary.ca

F. Samavati
e-mail: samavati@ucalgary.ca

M. Purss
Geoscience Australia, Canberra, Australia

X. Du
Key Laboratory of Digital Earth, Institute of Remote Sensing and Digital Earth, Chinese Academy of Sciences, Beijing, China

A. Mahdavi-Amiri (✉)
Department of Computing Science, Simon Fraser University, Burnaby, Canada
e-mail: amahdavi@sfu.ca

Two main approaches have arisen for the creation of a Digital Earth: Discrete Global Grid Systems (or DGGSs for short), and datacubes. In a DGGS, the surface of the Earth is discretized into a set of highly regular spherical/ellipsoidal cells. These cells are then addressed using a data structure or indexing mechanism that is used to assign and retrieve data. Datacubes are n-dimensional arrays that store geospatial data, ordered according to various attribute/coordinate axes, which can be spatial or non-spatial in nature.

In this chapter, we focus particularly on Discrete Global Grid Systems, which have global scope, more readily support interoperability than datacubes, are generally compatible with conventional datacube approaches, and can be used to provide the back-end support for a datacube implementation (Purss et al. 2019). The following section discusses DGGS as well as its components and their characteristics in detail.

2.2 Discrete Global Grid Systems

The traditional approach to discretizing the Earth is to use a latitude/longitude coordinate system on a sphere (Cozzi and Ring 2011), in which the 2D domain (or planar map of the Earth) is partitioned into a grid of cells by discretizing the 2D latitude/longitude domain. These cells may be further subdivided to increase the resolution and mapped to the sphere through the use of spherical coordinate equations and/or an appropriate spatial projection. The resulting spherical cells are primarily quadrilateral, though singularities and triangular cells appear at the poles, and the areas of the cells vary according to the latitude.

In order to better represent the Earth with more uniform cell structures, a polyhedron with the same topology as the sphere can be used. With an initial discretization of the Earth into planar cells (typically produced by considering the planar faces of an approximating polyhedron), the initial cells may then be refined to an arbitrary resolution and mapped from planar cells to spherical cells via some projection method (see Fig. 2.1). Given a regular refinement, a multiresolution hierarchy between cells (in which each cell has a coarser resolution parent and a number of finer resolution

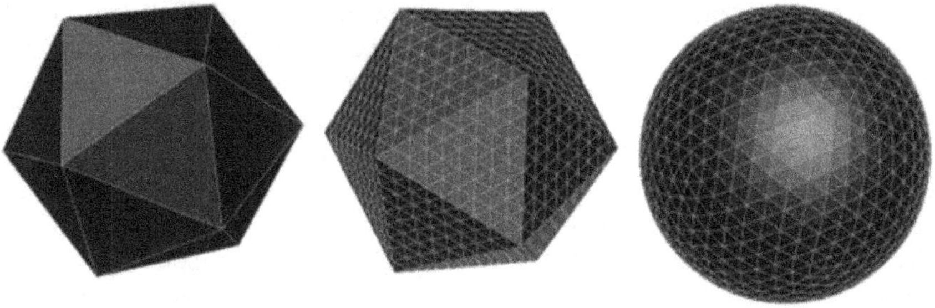

Fig. 2.1 An initial polyhedron that has been refined and projected onto the sphere

children) with a semiregular cell structure can be systematically defined. To query and render cells, they are then typically addressed via an indexing mechanism.

DGGSs are defined in terms of several different components. The main components of a DGGS include the initial planar or piecewise domain, cell type, projection, indexing, and refinement. In the following, we discuss each of these components in more detail in the context of DGGS construction.

2.2.1 Initial Domain

The simplest domain for a DGGS is a 2D map of the Earth, of which latitude/longitude grids are the standard. As these tend to exhibit large distortions across the globe and singularities at the poles, complicating queries and data analysis, a spherical polyhedron can instead be used for the initial domain of a DGGS. Such DGGSs are also known as Geodesic DGGSs (Sahr et al. 2003).

The most common choices for an initial polyhedron are the platonic solids—the tetrahedron, cube, octahedron, dodecahedron, and icosahedron—in addition to the truncated icosahedron. Each of these polyhedra offers distinct benefits. For instance, the octahedron defines a simple and symmetric domain, and can be unfolded into a very simple quadrilateral domain. Cubes are made of quadrilateral facets that are appropriate for the generation of quadrilateral cells. In comparison to other polyhedra, the icosahedron and truncated icosahedron undergo less angular distortion when processed through an equal area projection (Snyder 1992).

2.2.2 Cell Type

There are three main cell types that are used in a DGGS: hexagonal, triangular, and quadrilateral. If the initial domain is a polyhedron, other extraordinary cell types may be present, such as pentagons in a truncated icosahedron. However, the number of extraordinary cells is fixed no matter the resolution. Each of the three main cell types presents some advantages over the others that ought to be considered when selecting the base cell of a DGGS. For instance, quadrilateral cells are congruent, compatible with Cartesian coordinate systems, easy to index, and compatible with standard rendering libraries. Triangular cells are planar, easy to use, compatible with standard rendering techniques and libraries, and congruent. Hexagonal cells are the best for sampling, with the smallest quantization error, and they have uniform adjacency. Depending on the initial domain of a DGGS and the application that the DGGS was designed for, any of these types of cells can be employed as the DGGS cell type.

2.2.3 *Refinement*

Refinements are used to produce finer cells from an initial set of coarse cells. In a DGGS, they can be used to construct cells at multiple resolutions on the sphere by refining the faces of a polyhedron. Refinements are in part characterized by a ratio known as the aperture, or factor, of the refinement. This ratio relates the number of coarse cells to the number of fine cells at the next resolution, and several different apertures have been employed in DGGSs. After applying a refinement, the resulting fine cells may be assigned to a coarse *parent* cell as *children* of that cell, producing a hierarchical structure that is useful for many grid and spatial processing operations. Traversing from a parent cell to its children or from a child to its parent is known as hierarchical traversal.

In addition to the factor of refinement, other aspects play an important role in characterizing a refinement, such as congruency and alignment. In a congruent refinement, a coarse cell encompasses precisely the same area as a union of finer cells at the next resolution (see Fig. 2.2). For example, the 1-to-4 quadrilateral refinement shown in Fig. 2.2a is congruent while the 1-to-3 hexagonal refinement shown in Fig. 2.3b is not. Assigning a set of fine cells to serve as the children of a coarse cell is trivial in congruent refinements, as children are uniquely covered by a single parent cell and, therefore, the handling of hierarchical traversal queries is simplified. In contrast,

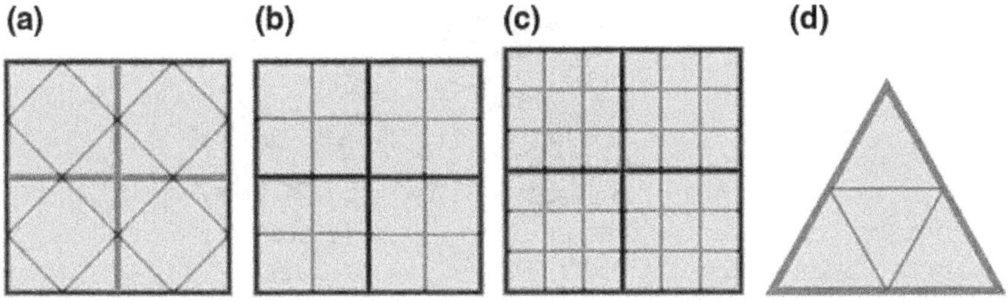

Fig. 2.2 (**a**) A center-aligned 1-to-2 refinement that is not congruent. (**b**) A vertex-aligned and congruent 1-to-4 refinement. (**c**) A center-aligned and congruent 1-to-9 refinement. (**d**) A center-aligned and congruent 1-to-4 refinement for triangles. The boundaries of coarser cells are highlighted using thicker lines

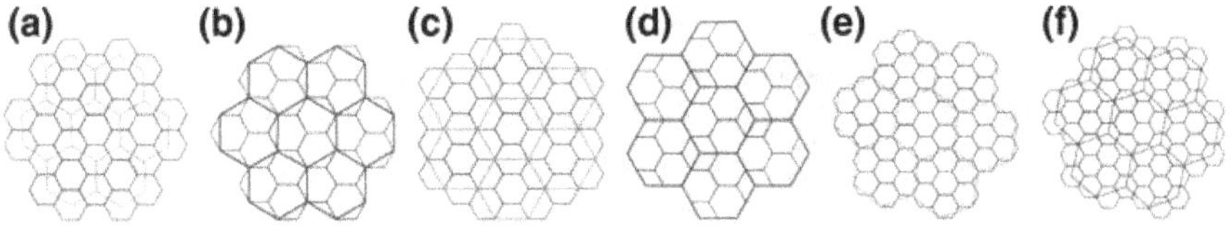

Fig. 2.3 (**a**) Center-aligned and (**b**) vertex-aligned 1-to-3 refinements. (**c**) Center-aligned and (**d**) vertex-aligned 1-to-4 refinements. (**e**) Center-aligned and (**f**) vertex-aligned 1-to-7 refinements. Note that each of these refinements is incongruent

incongruent refinements create ambiguity in the assignment of a child to a coarse cell, since there may exist several potential parents for a fine cell (Fig. 2.3).

If, after applying a refinement, every coarse cell shares its centroid with a fine cell, then that refinement is called center-aligned. Any DGGSs that employ such refinements are likewise called center-aligned. If such a property does not hold for the refinement, then it is usually vertex-aligned, meaning that the parent and child cells share a vertex (Fig. 2.2).

Various types of refinements exist for quadrilateral, triangular, and hexagonal DGGS cells. However, whereas many different refinements have been employed in DGGSs based on quadrilateral and hexagonal cells, DGGSs represented using triangular cells generally use 1-to-4 refinements (see Fig. 2.2d). Quadrilateral refinements can be congruent whilst hexagonal refinements never are (see Fig. 2.3). Consequently, parent-child relationships must always be explicitly defined in a hexagonal DGGS. However, once defined, this becomes a static feature of the DGGS infrastructure, allowing for consistent hierarchical traversal of the grid system, and incongruent refinements still exhibit characteristics that can be useful in a DGGS. Hexagonal 1-to-3 refinement has the lowest aperture of all hexagonal refinements; while hexagonal 1-to-4 refinement produces rotation-free lattices at all levels of resolution (simplifying hierarchical analysis in contrast to other refinements). Of the refinements shown in Fig. 2.3, fine cells in the 1-to-7 refinement cover the hexagonal coarse shape better than other refinements, and therefore more closely resemble congruency and provide a simpler hierarchical relationship between the cells. As a result, there is growing interest in this type of refinement (Middleton and Sivaswamy 2005).

2.2.4 *Projection*

Projections have traditionally been used to create maps of the Earth. Various forms of projection can be used to flatten the Earth (usually treated as spherical), and these can be categorized into different types, such as conformal, gnomonic, or equal area (Grafarend et al. 2014). When a spherical projection is used, some unavoidable distortions appear that one may try to reduce. In the following, we discuss several spherical projections in more detail.

2.2.4.1 Traditional Cartography

In traditional cartography, a spherical projection is a transformation from a point on the Earth (a sphere) to a point on a 2D map. Such a projection can be represented as a function: $P' = F^{-1}(P)$, where P lies on the sphere, and P' lies on the 2D map (see Fig. 2.4). The simplest method that can be used to create a 2D map from a spherical representation of the Earth is to use spherical coordinates $((\theta, \varphi))$. Doing so involves cutting the Earth (e.g. along a meridian) and unfolding it to form a 2D square, with θ (the longitude) and φ (the latitude) serving as the two main axes of

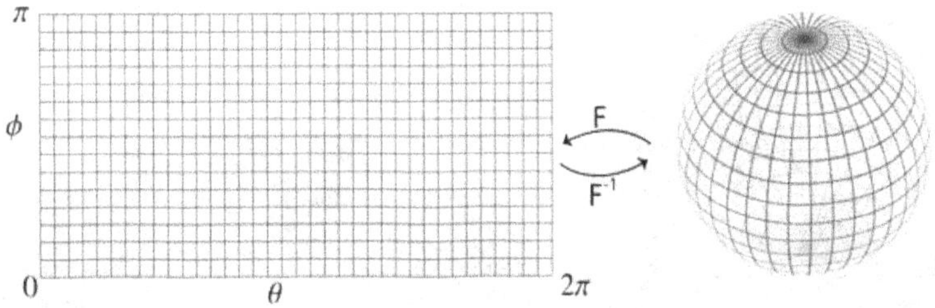

Fig. 2.4 2D domain and its corresponding sphere

the 2D domain. This 2D domain and its corresponding sphere are related through the following equations:

$$F(\theta, \varphi) = \begin{pmatrix} Rcos(\theta)\sin(\varphi) \\ Rsin(\theta)\sin(\varphi) \\ Rcos(\theta) \end{pmatrix}$$

$$F^{-1}(x, y, z) = \begin{pmatrix} \theta \\ \varphi \end{pmatrix} = \begin{pmatrix} tan^{-1}(\frac{y}{x}) \\ cos^{-1}(\frac{z}{R}) \end{pmatrix}$$

where R is the radius of the sphere: $R = \sqrt{x^2 + y^2 + z^2}$.

As can be observed from Fig. 2.4, perfect squares on the 2D domain are mapped to distorted quadrilateral or even triangular cells (at the poles) on the sphere. In order to reduce different distortions, it is possible to define alternative mappings, which can be equal-area, conformal (angle-preserving), or possibly stretch-preserving. Since the purpose of a DGGS is to use planar or piece-wise planar (i.e. polyhedral) domains to sample the spherical surface of the Earth, the use of an equal-area projection (in which data values sampled from the Earth occupy similar areas in the DGGS) is often desired (White et al. 1992). Here, we describe some of the most commonly used equal-area projections. There are several traditional options, such as the projections proposed by Lambert (cylindrical and azimuthal), Mollweide, and Werner (see Fig. 2.5) (Snyder 1992).

Among these projections, the Lambert azimuthal equal-area projection is particularly interesting, since it serves as the base projection for a number of equal-area polyhedral projections. In this projection, a point P on the sphere S is projected to a point P' on the 2D domain (2D map). To find P', a plane ρ which is tangent to S at another point C is used. Then, P' is the intersection of ρ with a circle that has its origin at C, passes through P, and is perpendicular to ρ (see Fig. 2.6). Note that the antipode of C (i.e. the point diametrically opposite C) is excluded from the projection as its intersecting circle is not unique. In addition, C is projected to itself along a circle of radius 0. This projection and its inverse can be explicitly described using simple mappings:

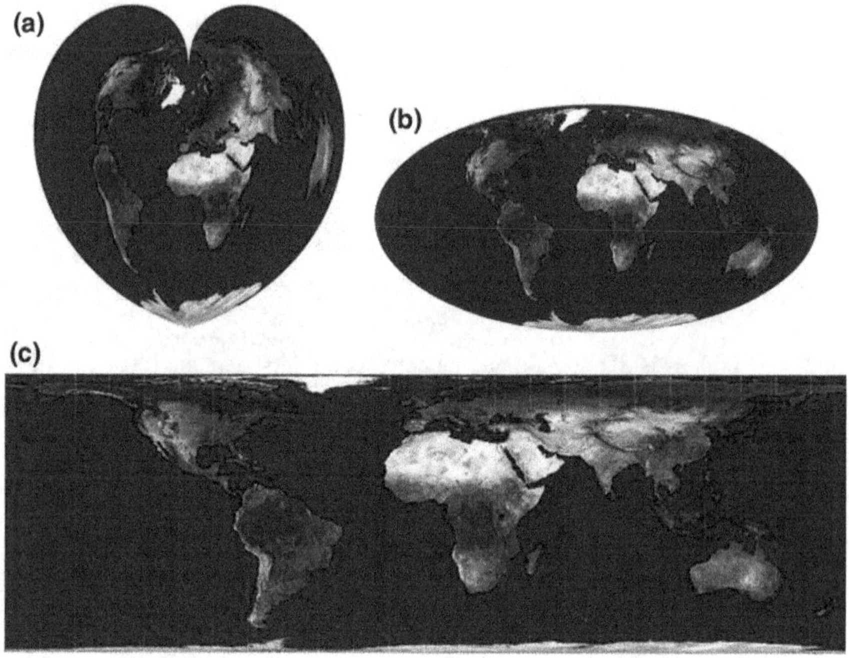

Fig. 2.5 (**a**) Werner, (**b**) Mollweide, and (**c**) Lambert (cylindrical) projections

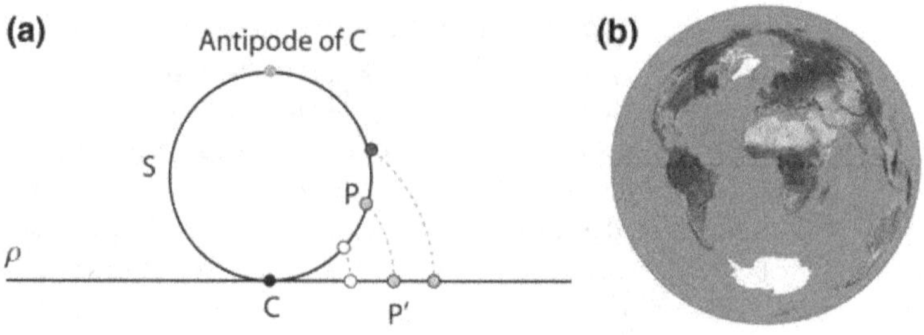

Fig. 2.6 (**a**) Lambert projection from sphere S to plane ρ. (**b**) 2D domain of the Earth resulting from Lambert Azimuthal equal-area projection. The second image is taken from Wikipedia

$$F(x, y, z) = \left(\sqrt{\frac{2}{1-z}}x, \sqrt{\frac{2}{1-z}}y\right),$$

$$F^{-1} = \left(\sqrt{1 - \frac{X^2+Y^2}{4}}X, \sqrt{1 - \frac{X^2+Y^2}{4}}Y, -1 + \frac{X^2+Y^2}{2}\right).$$

2.2.4.2 Projection for Polyhedral Globes

A projection defined for a given polyhedron is characterized by a function F that maps each point P on the sphere to a face of the polyhedron. Consequently F^{-1} is defined as a function that maps points on the polyhedron to the sphere. Often, in order to construct these projections, a traditional projection defined between the sphere and a plane is used individually on each face. For instance, Snyder's equal-area projection (which is commonly used in DGGSs) employs Lambert azimuthal equal-area projection individually for each face. The problem is reduced to projection for right triangles by splitting the faces of the polyhedron along lines of symmetry (see Fig. 2.7). A scaling factor is then found between the radii of two spheres: one with the same area as the (spherical) polyhedron, and another that circumscribes the polyhedron. Finally, a triangle on the polyhedron (whose area matches that of the corresponding spherical triangle that encompasses point P) is generated. The final equation for Snyder's projection is presented in closed form as:

$$x = \rho \sin(Az'),$$

$$y = \rho \cos(Az'),$$

in which the ratio ρ and angle Az' are defined in terms of a set of trigonometric functions and known constants that are provided in (Snyder 1992).

While the forward form of the Snyder projection is presented in a simple closed form, its inverse calculation requires finding the roots of a nonlinear equation (Snyder 1992). Snyder suggests the use of the Newton-Raphson iterative technique to compute the inverse projection, which can slow down the process of mapping points from the polyhedron to the sphere. To reduce this inefficiency, Harrison et al. (Harrison et al. 2011, 2012) worked to optimize the inverse Snyder projection by providing initial estimates to the Newton-Raphson technique that are close to the roots of the nonlinear

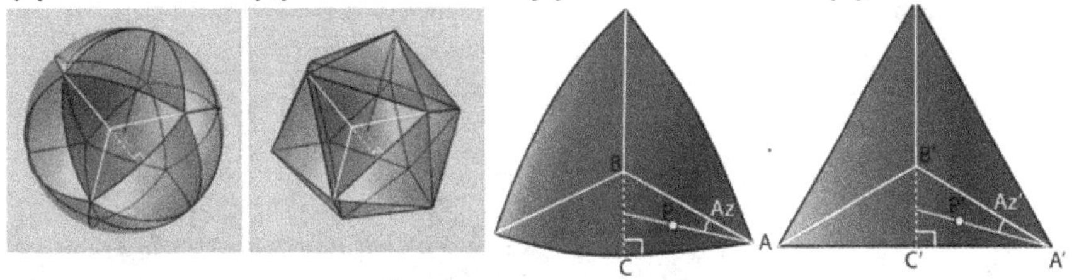

Fig. 2.7 An icosahedron (**a**) after and (**b**) before projection to the sphere. (**c**) The projection operates on right triangles by splitting the initial triangles. Each spherical triangle is associated with a triangle of the polyhedron. Each point P in (**c**) is associated with some point p̃ in (**d**). Az' is the angle between A and the great circle arc passing through P and A

equation. These initial estimates are found using a polynomial curve that fits the roots of the non-linear equation.

In addition to Snyder's projection, other types of equal-area projection have also been used in DGGSs. For instance, Roşca and Plonka's projection (Roşca and Plonka 2011) was used in the one-to-two Digital Earth, which is a cube-based Digital Earth (Mahdavi-Amiri et al. 2013), and extended to the octahedron in (Roşca and Plonka 2012). This equal-area projection describes a mapping from a cubic domain Ω to a spherical domain Δ. The main idea behind this projection is to map each face f of Ω onto a partition of Δ. To this end, an intermediate domain, called a curved square, is used. As a result, the projection involves two steps. First, f is mapped to a curved square on the tangent plane of Δ, parallel to f, using an equal-area bijection denoted by T. Then, the curved square is mapped to a partition of Δ using inverse Lambert azimuthal equal-area projection, denoted by F^{-1} (Fig. 2.8).

HEALPix (see Fig. 2.9) is also a cube-based equal-area projection, and is a hybrid projection of Lambert cylindrical equal-area projection and Collignon pseudo-cylindrical equal-area projection (Gorski et al. 2005). Lambert cylindrical projection is used for equatorial regions of the Earth while Collignon is used for the polar regions.

These equal-area projections are not the only projections that can be used in a DGGS but are examples of projections that have been used already. Naturally, a DGGS designer should always use a projection suited to the needs of their application.

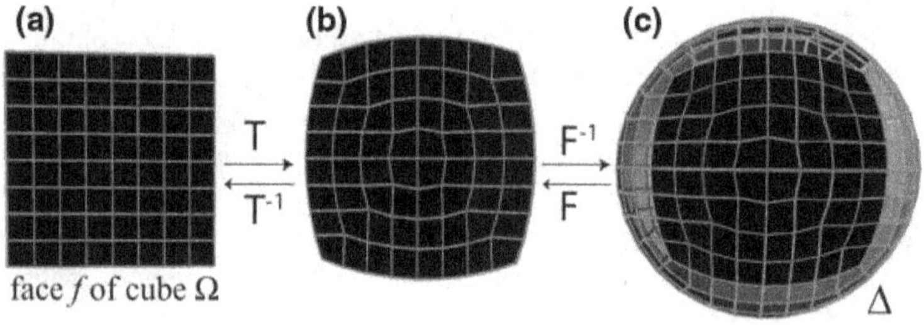

Fig. 2.8 Steps of the spherical projection. Points on a face f of Ω (**a**) are projected onto a curved square (**b**) and then projected onto a portion of the unit sphere Δ (**c**)

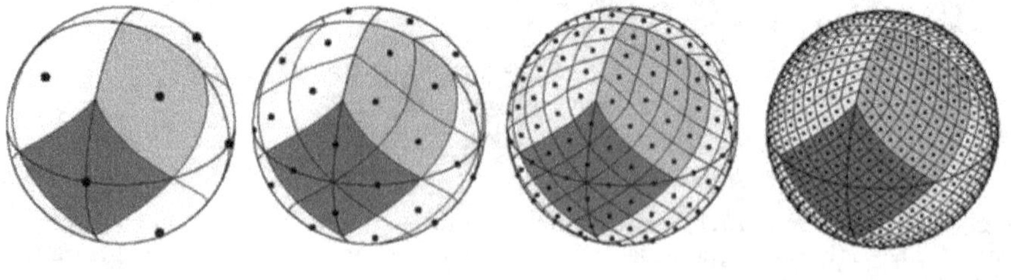

Fig. 2.9 HEALPix projection at four successive resolutions

2.2.5 *Indexing*

In order for a Digital Earth to handle queries related to location-based data efficiently, a hierarchical approach to data storage is needed. Hierarchical data structures such as quadtrees have been used in various Digital Earth frameworks (Fekete and Treinish 1990; Tobler and Chen 1986), but are typically shelved in favor of indexing methods in order to avoid the cost of expensive tree structures that record node dependencies. Given an indexing method for a DGGS, the method must ensure that, at each resolution, each cell receives an index that uniquely identifies the cell. This index may then be used with reference to a data structure or database in order to retrieve data associated with the cell.

Although various types of methods exist to index the cells of a DGGS, they are typically derived from three types of general indexing mechanisms: hierarchy-based, space-filling curve-based, and axes-based. In the following, we describe each category and provide some examples.

2.2.5.1 Hierarchy-Based Indexing

Applying refinements to a polyhedron produces a hierarchy that can be used to index cells. When a refinement is applied to a set of coarse cells, fine cells are created and assigned to coarse cells through a parent-child relationship. It is possible to use this relationship to define an indexing system by assigning an initial index to each cell at the first (i.e. lowest) resolution, and then using each cell's index as a prefix to the indices of its children. Formally, if a coarse cell has index $I d_0 d_1 \ldots d_{r-1}$, then its children receive indices of the form $I d_0 d_1 \ldots d_{r-1} d_r$, where d_r is an integer whose range is known as the base of the indexing method, denoted by b (i.e. $d_i \in [0, b-1]$).

The base is used to define algebraic operations on indices, such as conversion to and from the Cartesian coordinate system, or neighborhood finding (Tobler and Chen 1986; Vince and Zheng 2009). Hierarchy-based indexing is very efficient in supporting hierarchical queries, although neighborhood-finding tasks may require complex algorithms, depending on the base of the indexing method and the algebra defined for the indexing system. An example of this type of indexing was proposed in (Gargantin 1982) for quadrilateral cells resulting from 1-to-4 refinement (see Fig. 2.10). Here, the children resulting from 1-to-4 refinement on a quadrilateral with index I receive indices I_0, I_1, I_2, I_3.

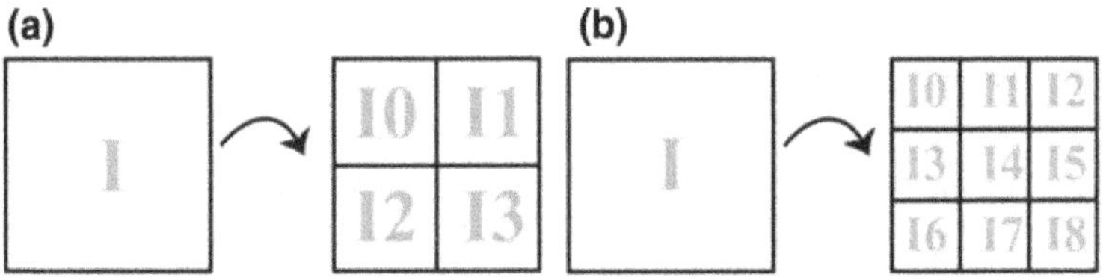

Fig. 2.10 Hierarchy-based indexing systems in (**a**) base four and (**b**) base nine

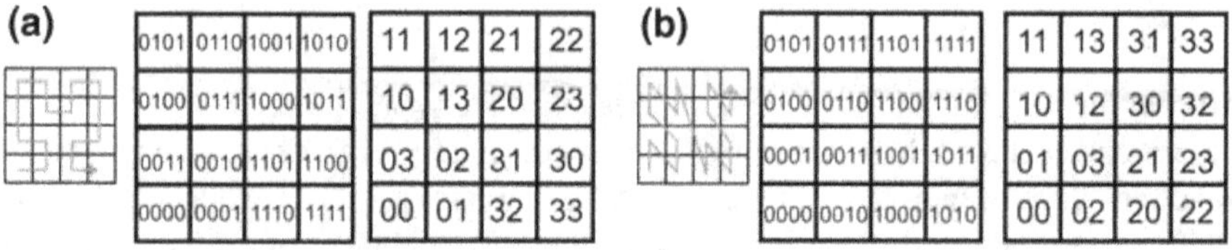

Fig. 2.11 (**a**) (Left) Hilbert SFC. (Middle) Indexing in base two. (Right) Indexing in base four. (**b**) (Left) Morton SFC. (Middle) Indexing in base two. (Right) Indexing in base four

2.2.5.2 Space-Filling Curve Indexing

Another method for indexing cells in a DGGS is to use a space-filling curve (SFC) as a reference for the indexing (Mahdavi-Amiri et al. 2015b). SFCs have been used in many applications, such as compression, rendering, and database management, and are 1D curves (often recursively defined) that cover a particular space.

Recursively defined SFCs start from a simple initial geometry defined on a simple domain (usually a square). The domain is then refined and the simple geometry is repetitively transformed to cover the entire refined domain. Typically, if the initial geometry covers i cells, a 1-to-i refinement is suitable for use in generating a refined domain. In this way, each SFC is associated with a refinement. To index cells based on an SFC, decimal numbers can be employed, though the corresponding indices do not directly encode the resolution. To resolve this issue and obtain a compact indexing, a base b for the indexing that is compatible with the refinement is often chosen. With a 1-to-i refinement, this usually means that the base of the indexing method is taken to be i or $\sqrt{i}$ (see Fig. 2.11). For instance, the refinement associated with the Hilbert and Morton curves is 1-to-4. Therefore, when using a Hilbert and Morton curve, a base four or base two indexing is appropriate.

Indexing methods derived from SFCs have been widely used in DGGS and terrain rendering. For instance, in (Bai et al. 2005; White 2000), Morton indexing was used to index cells resulting from 1-to-4 refinements on the icosahedron and octahedron, while in (Bartholdi III and Goldsman 2001), the Sierpinski SFC was used to index triangular cells refined with a factor of two.

2.2.5.3 Axes-Based Indexing

Another mechanism for indexing is to use a coordinate system with m axes, U_1 to U_m, that span the entire space on which the cells lie. Then the cells' indices can be expressed as m-dimensional vectors $(i_1, i_2, \ldots, i_m)$, in which the i_j are integer values that indicate the number of unit steps taken along each axis, U_j. If, alternatively, a 1D

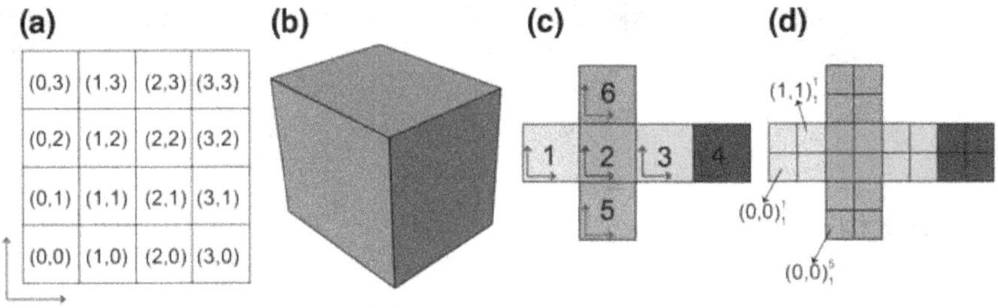

Fig. 2.12 (**a**) Integer indexing of cells using Cartesian coordinates. (**b**) A cube. (**c**) The unfolded cube in (**b**) and coordinate systems for each face. **d** Indices of some cells after one step of 1-to-4 refinement

index is required, these integer values can be appended together in a string, separated using a delimiter character. A simple example of such an indexing can be used to index a quadrilateral domain with Cartesian coordinates, as illustrated in Fig. 2.12a. When a refinement is applied to the cells, a subscript r is appended to the index in order to encode the resolution. In the proposed axes-based indexing methods for DGGSs, m is typically taken to be two or three. For instance, 3D indexing was used in (Vince 2006) by taking the barycentric coordinate of each cell to be its index.

A 2D indexing method can be applied on the polyhedron used to construct a DGGS by embedding the polyhedron's faces into a 2D domain and defining a coordinate system over that domain (Mahdavi-Amiri and Samavati 2012). In this way, each face can be given its own coordinate system (Mahdavi-Amiri et al. 2013, 2015a; Mahdavi-Amiri and Samavati 2014). Figure 2.12 illustrates an indexing for the quadrilateral cells of a cube after 1-to-4 refinement, where each face has its own coordinate system. In order to distinguish between the cells associated with each face, an additional component that refers to the initial polyhedral face can be added to the indices. For example, index $(a, b)_r^f$ refers to cell (a, b) in face f at resolution r (Fig. 2.12d).

2.2.5.4 Remarks on Categorization

Note that this categorization of index types is primarily intended to reflect the core idea used to construct the indexing methods and can be used to easily identify which operations can be handled naturally by a particular indexing system. For example, hierarchy-based indexing methods naturally lead to efficient hierarchical traversal operations. However, well-designed indexing methods must necessarily also consider other properties and support other operations. For example, it is certainly possible to handle neighborhood finding in hierarchy-based indexing methods, although not as efficiently or as naturally as with axes-based techniques. Based on the pattern of indices, some indexing methods can be interpreted as belonging to two categories (e.g. SFC or hierarchy-based). However, an indexing method is either constructed through use of a parametrized curve or it inherits the index of its parent. It is possible to use a parametrized SFC that indexes the children and prefixes the parent's index.

This indexing method is considered to be SFC-based, since the construction of the indexing is based on the parametrization of the SFC and not on the hierarchy of the cells.

2.3 Scientific Digital Earths

Now that we have discussed the various components that define different DGGSs, let us examine some specific DGGS constructions that have been proposed in the literature. Note that some proposed DGGSs are left for the following section, in which we survey some of the existing Digital Earth implementations.

Among the earliest proposed DGGSs are those designed by Digital Earth pioneers M. Goodchild and G. Dutton. Goodchild's HSDS (Hierarchical Spatial Data Structure (Goodchild and Shiren 1992)) is built upon an initial octahedron. Unlike many other DGGSs, the faces of this octahedron are inverse projected to the sphere before the generation of finer cells, and the refinement (a congruent, 1-to-4 refinement) is applied directly on the resulting spherical triangles (using geodesic rather than Euclidean midpoints). The child cells are also spherical triangles, and are indexed using a hierarchical base-4 numbering scheme (see Fig. 2.13). Unfortunately, the projection implied by the refinement method is neither equal-area nor conformal.

Dutton's QTM (Quaternary Triangle Mesh (Dutton 1999)) is also constructed using congruent 1-to-4 refinement on an initial octahedron but utilizes the more typical refine-then-project approach (see Fig. 2.14). Here, the employed projection is a specially designed projection that is also neither equal-area nor conformal—the ZOT (Zenithal Orthotriangular Projection (Dutton 1991))—which tries to have similar facets with vertices spaced uniformly in latitude and longitude, as well as low areal distortion. Indexing is performed similarly to the HSDS, with the faces of the initial octahedron indexed 1 through 8, and child cells indexed hierarchically in base 4.

SCENZ-Grid (SEEGrid 2019) is a DGGS that is constructed based on an initial cube polyhedron, and which was created through a collaboration between Landcare Research and GNS Science primarily for the purpose of environmental monitoring.

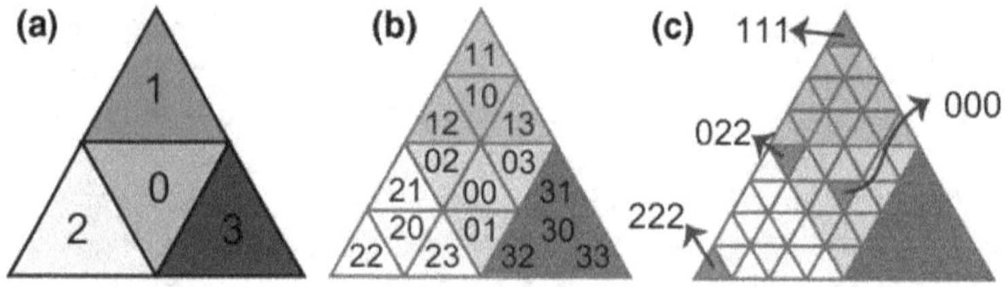

Fig. 2.13 (**a, b**) The hierarchical indexing system of the HSDS. (**c**) Indices of descendant cells after three refinements

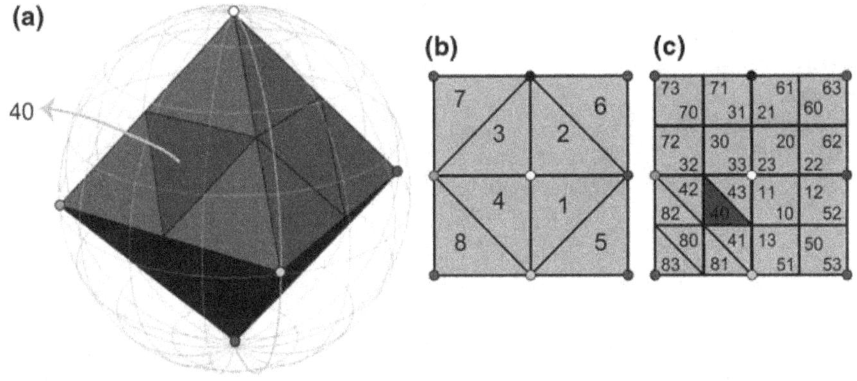

Fig. 2.14 The QTM DGGS. (**a**) The initial octahedron, embedded in a sphere. (**b**, **c**) The hierarchical indexing system of the QTM

The faces of the initial cube are refined using a 1-to-9 congruent and aligned quadrilateral refinement, and the resulting cells are inverse projected using the HEALPix projection method (see Sect. 2.2.4.2). A hierarchical base-9 indexing system is used to address the cells (see Fig. 2.15).

Quadrilateral cells are also found in Crusta (Bernardin et al. 2011), a DGGS based on a rhombic triacontahedron. Each of the initial 30 quadrilateral faces undergoes a 1-to-4 refinement, and the generated vertices are normalized to the geoid. Crusta's primary motivation includes support for high-resolution topographical data and images.

A number of hexagon-based DGGSs have also been proposed and have garnered much research attention. The ISEA3H (Icosahedral Snyder Equal Area Aperture 3 Hexagonal) DGGS is a particularly notable example which starts from an icosahedron (or truncated icosahedron) that undergoes an aligned 1-to-3 hexagonal refinement (US Patent No. 8400451, 2004; Sahr 2008). The resulting cells are inverse projected to the sphere using Snyder's equal-area projection. Note that as the refinement scheme

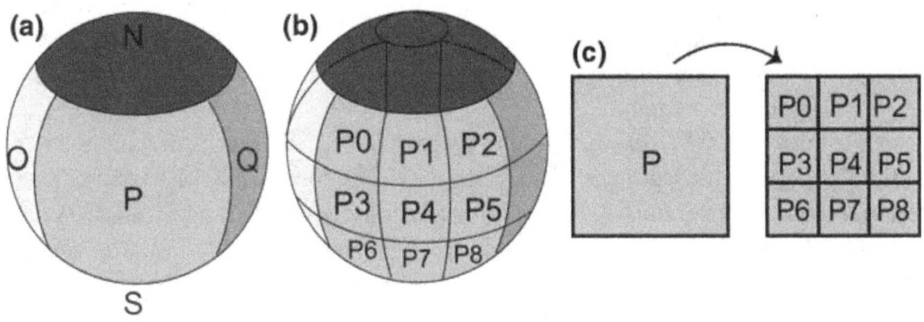

Fig. 2.15 (**a**), (**b**) SCENZ-Grid is created from a refined cube inverse projected to the sphere. (**c**) The hierarchical indexing system of SCENZ-Grid

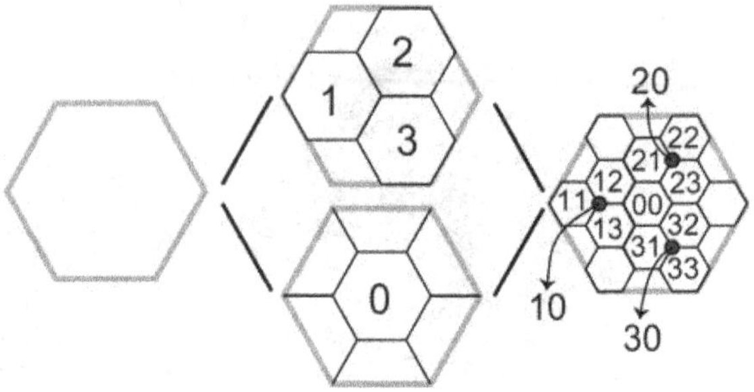

Fig. 2.16 Two types of 1-to-4 hexagonal refinement can be combined, allowing a triangular cell hierarchy to be established and indexed

(and, indeed, any hexagonal refinement scheme) is not congruent, special care must be taken to define the cell hierarchy and indexing scheme. Several different indexing schemes have been proposed for the ISEA3H DGGS. These include the hierarchical indexing of PYXIS (US Patent No. 8400451, Peterson 2004), CPI (US Patent No. 9311350, Sahr 2016), coordinate-based indexing mechanisms (Sahr 2008; Mahdavi-Amiri et al. 2015a; Vince 2006), or the algebraic encoding scheme of (Ben et al. 2018).

The icosahedron can also be refined using 1-to-4 hexagonal refinement, as in the construction of the HQBS (Hexagonal Quaternary Balanced Structure (Tong et al. 2013)). The resulting cells are also inverse projected using Snyder's projection. In order to mitigate the incongruity of the hexagonal refinement, two different 1-to-4 refinements are employed (aligned and unaligned; see Fig. 2.16). This allows a triangular hierarchy to be defined (aligned with the edges of the initial icosahedron) and for lattice points to be indexed. By taking the index of the point at the cell's centroid to be the cell's index, a base 4 hierarchical indexing system can be established on the cells.

Other hexagon-based DGGSs include the OA3HDGG and OA4HDGG (Octahedral Aperture 3/4 Hexagonal Discrete Global Grid (Vince 2006; Ben et al. 2010)). As implied by the name(s), both DGGSs are constructed from an octahedron that undergoes a hexagonal refinement. The OA3HDGG utilizes a 1-to-3 hexagonal refinement, and its cells are indexed using a coordinate-based system. The vertices of the initial octahedron are assigned the coordinates $(\pm 1, 0, 0)$, $(0, \pm 1, 0)$, and $(0, 0, \pm 1)$; and the cells are assigned indices based on their barycentric coordinates with respect to these vertices. A similar indexing system is applied to the OA4HDGG, which utilizes a 1-to-4 hexagonal refinement.

While most DGGSs discretize only the surface of the Earth, certain types of geospatial data (e.g. earthquake data, airspace delineations, etc.) are volumetric in nature and require a volumetric Earth representation. Hence, volumetric DGGSs such as SDOG (Spheroid Degenerated-Octree Grid (Yu and Wu 2009; Yu et al. 2012)) have also been proposed. SDOG, which was designed to represent the global lithosphere, divides the Earth into an initial set of eight octants. Each octant is associated with a degenerated octree, and undergoes a non-uniform refinement that prevents cell degeneracies at the Earth's core (see Fig. 2.17). Cells are indexed using two different

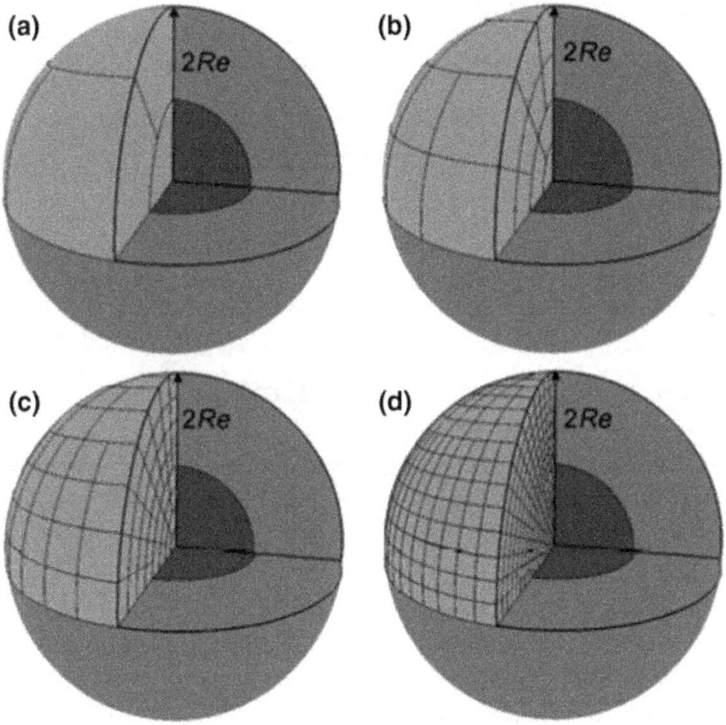

Fig. 2.17 The SDOG volumetric DGGS at three successive resolutions

curve-based schemes, both based on a modified Z-curve. The SDZ (Single Hierarchical Degenerated Z-Curve Filling) method indexes the cells of a single resolution in base 10 (see Fig. 2.18), while MDZ (Multiple Hierarchical Degenerated Z-Curve Filling) serves as a hierarchical indexing scheme in base 8 (see Fig. 2.19).

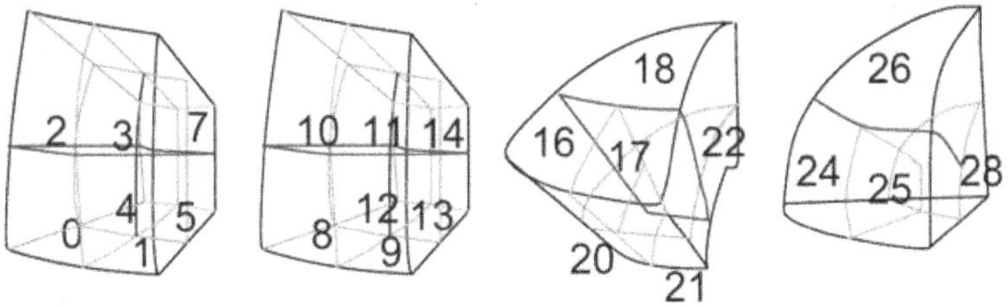

Fig. 2.18 SDZ defines a base 10 indexing for cells at a single resolution

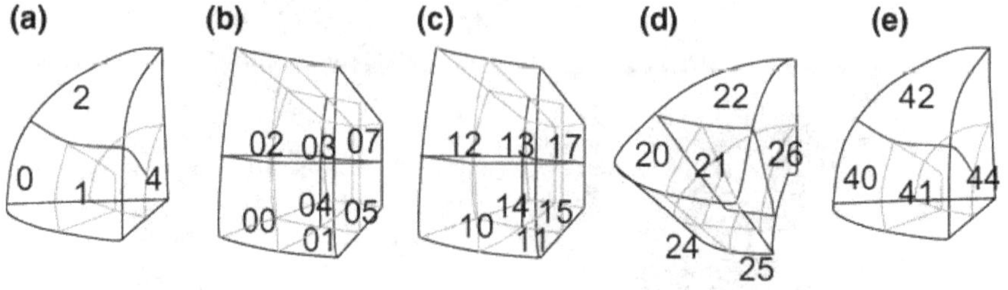

Fig. 2.19 MDZ defines a hierarchical indexing scheme on the SDOG cells. (**a**) A refined octant with child indices. (**b**), (**c**), (**d**) and (**e**) Cells 0, 1, 2, and 4, respectively, refined with child indices

2.4 Public and Commercial Digital Earth Platforms

Naturally, a number of DGGSs and other Digital Earth concepts have been implemented and made available for public use as either free or paid software.

2.4.1 Latitude/Longitude Grids

Due to their ease of use and long history, latitude/longitude grids remain a popular choice for Digital Earth implementations despite the potential issues associated with non-uniform DGGS cells. Chief among these implementations in terms of name recognition is Google Earth (Google Inc. 2019a). Google Earth is created upon a latitude/longitude grid using a simple cylindrical projection, with textures processed via clip-mapping (Bar-Zeev 2007). Clip-Maps are a modified form of mip-map that impose a maximum image size on the mip-map hierarchy (Tanner et al. 1998), causing the image hierarchy to more closely resemble an obelisk than the traditional pyramid (see Fig. 2.20). The capped image size ensures that textures can fit into memory and be rendered in real-time.

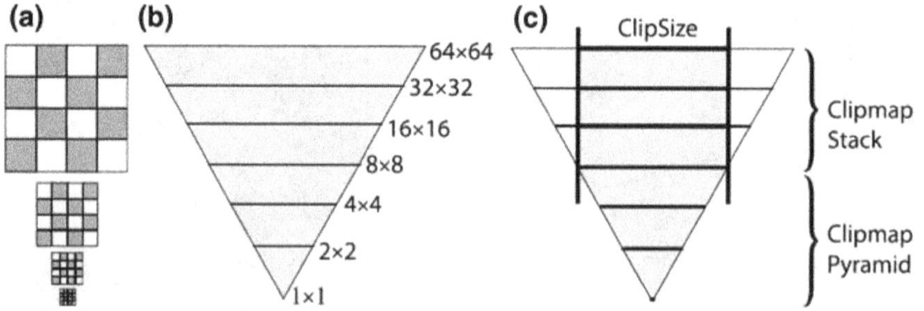

Fig. 2.20 (**a**) An image at multiple resolutions. (**b**) The image's mip-map pyramid. **c** Clip-Maps impose a maximum size on the mip-maps

Although not presented in 3D globe format, Google Maps and Bing Maps are supported using methods that echo the fundamentals of a DGGS (See Figs. 2.21 and 2.22). In particular, Google Maps uses the Mercator projection on a latitude/longitude grid that is refined using a 1-to-4 refinement. Each “tile” of the hierarchical grid is associated with a 256 × 256-sized texture, and is indexed using an axis-based coordinate system. Here, the top-left tile is indexed as (0, 0), with x values increasing towards the east, and y values increasing towards the south (Google Inc. 2019b).

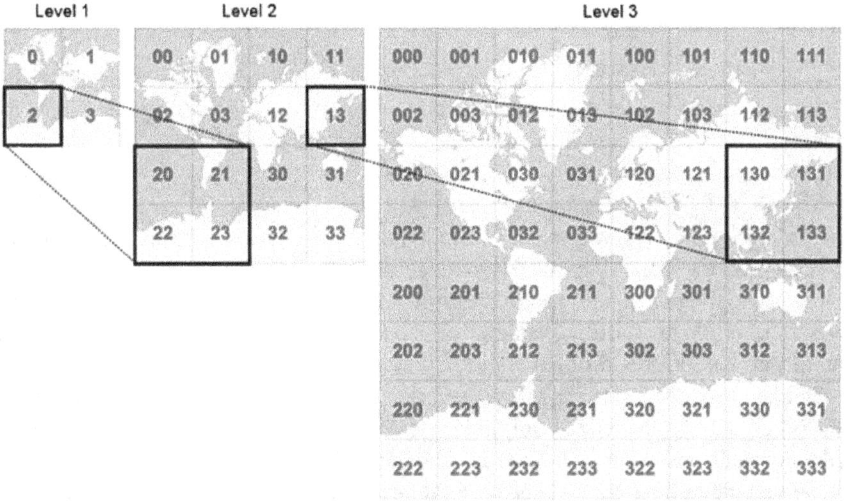

Fig. 2.21 Three resolutions of Bing Maps’ hierarchy-based indexing

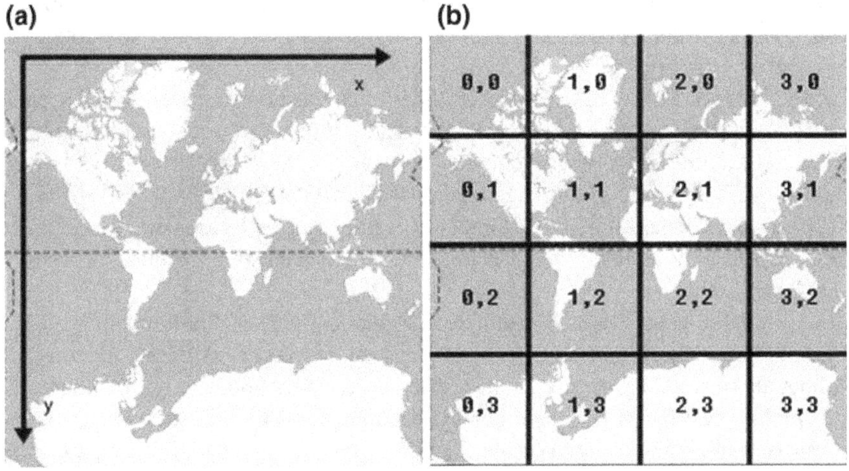

Fig. 2.22 (**a**) Google Maps’ latitude/longitude grid. (**b**) Cell indices at the second resolution

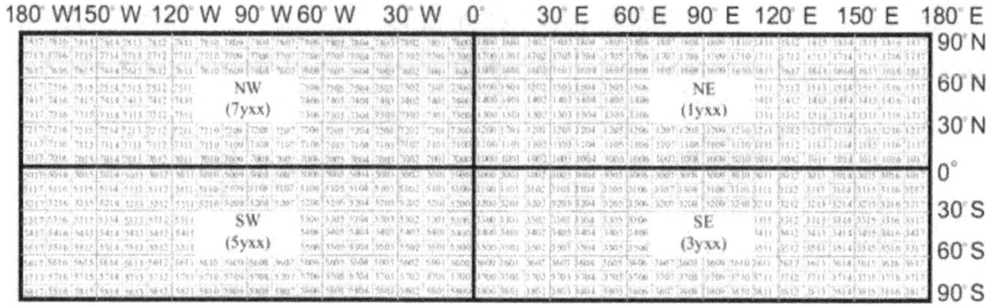

Fig. 2.23 C-squares indexing system

Bing Maps also uses the Mercator projection on a 1-to-4 refined grid, but its indexing system is hierarchical and based on quadtrees (Schwartz 2018). For an illustration, see Fig. 2.21.

The OGC CDB (Common Database) API from Presagis (2019) is designed to address one of the main issues with latitude/longitude DGGSs, namely the shrinking of cells near the poles. The CDB divides the Earth into five zones depending on proximity to the poles, with each zone utilizing a different spacing between lines of longitude. While the CDB is available as an open commercial standard, a Pro license can be purchased for additional features.

Unlike other DGGSs, C-squares (Concise Spatial Query and Representation System (Rees 2003)) discretizes only a single resolution of the Earth. Here, the latitude/longitude grid is divided into four quadrants (NE, NW, SE, and SW), which are then divided into finer grids based on latitude and longitude (Fig. 2.23). The cells of this discretization are indexed as *iyxx*, where *i* corresponds to the cell's quadrant, *y* to the cell's latitude, and *xx* to the cell's longitude. This system was created by CSIRO Marine and Atmospheric Research, Australia for the purposes of mapping, spatial search, and environmental assessment. Converters and source code can be found on their website (CSIRO 2019).

Other Digital Earths based upon latitude/longitude grids include NASA's open source WorldWind API (NASA 2019); Skyline's software products, TerraExplorer client and SkylineGlobe server (Skyline Software Systems 2019); and two DGGS libraries for web-based globe visualization—GlobWeb and CesiumJS (Telespazio 2019; Cesium Consortium 2019). GlobWeb is provided by Telespazio France under the GNU LGPL v3 license, while CesiumJS was founded by the Cesium Consortium and is open source.

CesiumJS in particular is a complete 3D mapping platform built using WebGL. It is a cross-platform and cross-browser map engine that runs on a web browser without plugins, and is now used in industries as diverse as archaeology, engineering, construction, and sports visualizations. An accompanying tool, Cesium ion, provides a point-and-click workflow to create 3D maps of users' geospatial data that can be visualized, analysed, and shared. It can be used to host datasets in 3D tiles, including imagery, terrain, photogrammetry, point clouds, BIM, CAD, 3D buildings, and vector data; and provides tools for analytics including measurements, volume and visibility computations, and terrain profiles.

2.4.2 Geodesic DGGSs

Of course, not all DGGSs are based on singular 2D domains such as a latitude/longitude grid; while comparatively rarer, different implementations of Geodesic DGGSs do exist and are available for use. For instance, a library that implements the well-studied ISEA3H DGGS—known as geogrid—is offered on GitHub (Mocnik 2019). This library is developed and maintained by Franz-Benjamin Mocnik, and is licensed under the MIT license.

A propriety implementation of the ISEA3H can be found at the core of the Digital Earth system developed by Global Grid Systems (formerly the PYXIS innovation (PYXIS innovation 2011; Global Grid Systems 2019)), and is one of the few commercially available Geodesic DGGSs (Fig. 2.24). This system indexes the ISEA3H's hexagonal cells using the patented PYXIS indexing scheme (US Patent No. 8400451, 2004).

Other software platforms include implementations of the ECM (Ellipsoidal Cube Map) and HEALPix (Hierarchical Equal Area isoLatitude Pixelization of the sphere) DGGSs. ECM (Lambers and Kolb 2012) is produced by applying 1-to-4 refinement on the quadrilateral faces of a cube that circumscribes the ellipsoidal Earth. Areal and angle distortions are minimized by using a Quadrilateralized Spherical Cube (QSC) projection. A Linux implementation is available on Martin Lambers' website, licensed under the GNU GPL v3 (Lambers 2019).

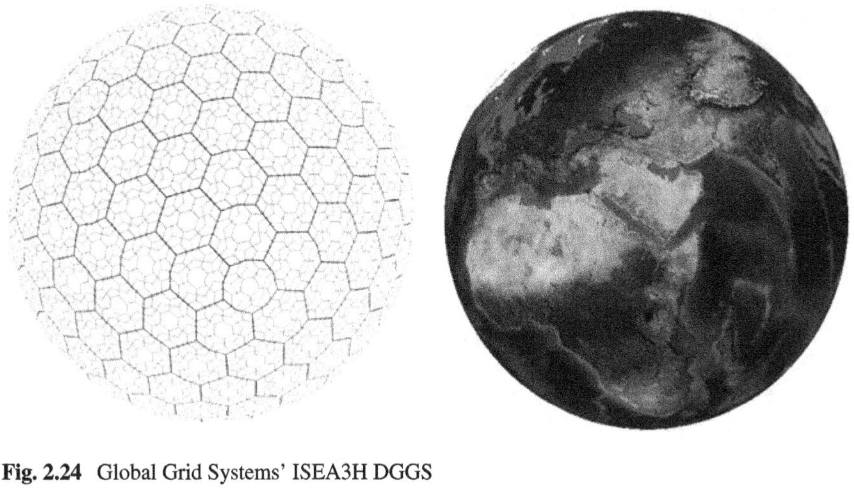

Fig. 2.24 Global Grid Systems' ISEA3H DGGS

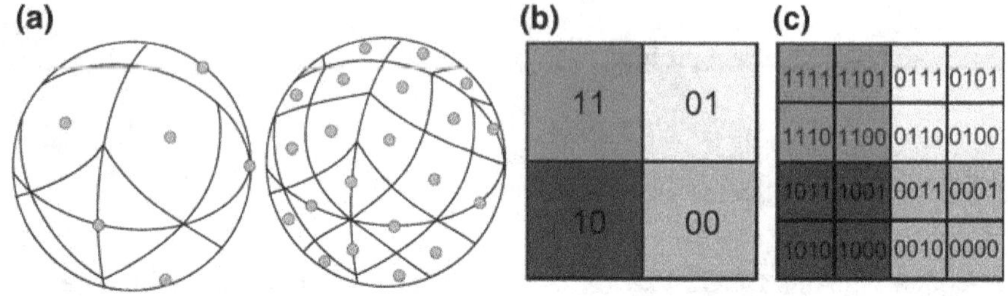

Fig. 2.25 The HEALPix DGGS. (**a**) HEALPix uses a 1-to-4 refinement. (**b**) The hierarchical HEALPix indexing system

The HEALPix DGGS (Gorski et al. 2005) is based on a rhombic dodecahedron that undergoes a congruent 1-to-4 refinement, indexed using a base-2 hierarchical indexing system (see Fig. 2.25). Two different projection schemes are employed: Lambert cylindrical equal-area projection for equatorial regions, and Collignon equal-area projection for polar regions. A software package from the Jet Propulsion Laboratory that supports spherical harmonics, pixel queries, data processing, and statistical analysis can be found online (Jet Propulsion Laboratory 2019).

2.4.3 Installations: DESP

One of the largest scale Digital Earth undertakings can be found at the Chinese Academy of Sciences (CAS), where an interactive visualization environment called the Digital Earth Science Platform was developed (Guo et al. 2017). Based on the Digital Earth Prototype System Initiative that launched in 1999 (Guo et al. 2009, 2010), the Digital Earth Science Platform (DESP) was established by the CAS in 2010 in order to integrate state-of-the-art techniques and meet the increasing requirements of geoscience applications.

The DESP is a technical platform that supports spatial data and information services, as well as associated applications. It integrates 2D and 3D geographic information systems, distributed storage and computing, virtual reality, wireless sensor networks, and other technologies. A 600 m^2 fully immersive, interactive visualization environment was established at the CAS Institute of Remote Sensing and Digital Earth (RADI) to support experiments with 3D visualization and to provide decision support for emergency response applications. This installation is equipped with VR/AR devices, sensors, a 3D Stereo Projection System, and a high-performance computing system, as shown in Fig. 2.26.

The DESP has already played an important role in the modeling of global change, evaluation of natural disasters, and monitoring of natural resources and human settlement through the integration of multi-sensor, multi-temporal remote sensing images, in situ ground survey data, socio-economic data, and interdisciplinary scientific models (Fan et al. 2009). For example, the influence of sea level rise on the Earth's major

Fig. 2.26 The DESP visualization environment at RADI

river deltas has been modeled and analyzed in a comparative study by using the DESP. Emergency monitoring and response systems based upon the DESP have been developed for disaster monitoring and post-disaster relief after earthquake events (Fig. 2.27).

As a part of the ongoing Big Earth Data Science Engineering (CASEarth) initiative (2018–2022), which is supported by the Strategic Priority Research Program of the CAS, a new generation of the Digital Earth Science Platform will be developed to provide a new impetus for interdisciplinary, cross-scale, macro-scientific discoveries in the Era of Big Data to promote sustainability (Guo 2017).

Fig. 2.27 The DESP was used for disaster assessment and relief after the 2013 Ya'an earthquake

2.5 Discrete Global Grid System Standards

The myriad ways in which one can construct a Digital Earth platform provide a great deal of flexibility that can help cater to a vast range of specific uses; however, this can also create barriers to interoperability. This creates a real challenge as we move into the Era of Big Data (and beyond), where interoperability and distributed analysis is critical.

In the Era of Big Data, geoscience can only achieve its full potential through the fusion of diverse Earth observation and socio-economic data together with information from a vast range of sources. In this type of environment with multiple data providers, fusion is only possible with an information system architecture based upon open standards (Percivall 2013). Without a common and standardized means of defining and integrating various Digital Earth Platforms, our ability to transform the increasingly massive amounts of data that are being acquired about the Earth into actionable information is significantly limited.

2.5.1 Standardization of Discrete Global Grid Systems

Recognizing the issues that non-standard global grid system implementations pose and their potential impacts, in 2014, the Open Geospatial Consortium embarked on the ambitious goal of standardizing DGGS. The goal of this endeavor was not to identify the one DGGS that ought to be used by everyone, but to define the common qualities of a variety of DGGSs that can be used to support interoperability while providing some flexibility in choice, thus allowing implementers to tailor DGGS infrastructures to their specific uses. In July 2017, the OGC published the first ever international standard governing the design and implementation of DGGS (Purss et al. 2017). This standard aims to promote awareness and reusability of DGGS implementations, and integration between them, and, through this, to demonstrate a path towards the realization of the "Digital Twin"—where our engagement and understanding of the physical Earth can seamlessly interact with the Digital Earth, and vice versa.

The core of the OGC DGGS standard is primarily based on an appropriate subset of the well-accepted criteria for optimal DGGS design proposed by Goodchild (2000) and Sahr et al. (2003).

2.5.2 Core Requirements of the OGC DGGS Abstract Specification

Along with the categorization provided earlier, under the OGC DGGS Abstract Specification, a compliant DGGS must define a hierarchical tessellation of equal area cells that both partition the entire Earth at multiple levels of granularity and provide a global spatial reference frame. In addition to these structural components, the system must also include encoding methods to address each cell, assign quantized data to cells, and perform algebraic operations on the cells and the data assigned to them.

The requirement of functional components for the infrastructure sets an OGC DGGS apart from other grid frameworks or Coordinate Reference Systems. It also provides a common operational basis for supporting communication and interoperability between different compliant DGGS infrastructures.

2.5.2.1 Structural Requirements

The reference frame of a DGGS consists of the fixed structural elements that define the spatial framework on which the DGGS functional algorithms operate. These fixed structural elements include:

1. **Domain completeness and position uniqueness**: The DGGS must be defined over a global domain without any overlapping cells. Goodchild defines a global domain to be achieved when the areal cells defined by the grid "*exhaustively cover the globe without overlapping or underlapping*" (Goodchild 2000);
2. **Multiple levels of resolution**: The DGGS must define multiple discrete global grids forming a system of hierarchical tessellations, each with progressively finer spatial resolution and linked via a common cell refinement method;
3. **Preservation of domain completeness and position uniqueness**: The DGGS must preserve the total surface area (i.e. the global domain) throughout the entire range of hierarchical tessellations. This facilitates the consistent representation of information at all resolutions within the DGGS;
4. **A simple geometric structure for each cell**: In order for the DGGS to achieve the requirement of a global domain, it is necessary for the shape of all cells defined by the DGGS to be simple polygons on the surface model of the Earth. The cell shapes derived from the five (5) Platonic solids and thirteen (13) Archimedean solids (triangle, quadrilateral, pentagon, hexagon, and octagon) are all simple polygons that have the following properties:

    a. The edges meet only at the vertices;
    b. Exactly two edges meet at each vertex; and,
    c. The polygons enclose a region which always has a measureable area.

5. **Equal-area cells**: The DGGS must be based on a hierarchy of equal-area tessellations. Equal-area cells provide global grids with spatial units that (at multiple resolutions) have an equal probability of contributing to an analysis. Equal-area cells also help minimize the confounding effects of area variations in spatial analyses, where the curved surface of the Earth is the fundamental reference frame;
6. **An initial polyhedral tessellation**: To consistently achieve equal-area cells, the DGGS must be constructed by mapping a polyhedron to the surface model of

the Earth. This initial tessellation can then be refined to produce equal-area child cells for all subsequent levels in the hierarchy of tessellations;

7. **Unique identifiers for each cell**: In order to efficiently operate as a spatial data integration engine, the cells of the DGGS must each be defined by a globally unique identifier. This ensures that the reference to each and every cell is immutable. While the OGC DGGS Abstract Specification requires each cell to be uniquely identified, it does not prescribe or enforce how the implementer must achieve this;
8. **Each cell referenced at its centroid location**: Each DGGS cell must be referenced at its centroid. This is because the centroid is the only location that provides a systematic and consistent spatial reference point for all cells, regardless of shape.

2.5.2.2 Functional Requirements

The ability to locate and perform algebraic operations on data assigned to a DGGS is critical for a DGGS to be able to support connectivity and hierarchical operations on cells and to facilitate interoperability between DGGS implementations (as well as other spatial data infrastructures or interfaces). Accordingly, the OGC DGGS Abstract Specification requires the DGGS to specify definitions for:

1. **Quantization operations**: Assigning data to and retrieving data from cells;
2. **Algebraic operations**: Performing algebraic operations on cells and the data assigned to them, in addition to performing cell navigation; and
3. **Interoperability operations**: Translating cell addresses to other Coordinate Reference Systems (CRS), such as conventional latitude/longitude.

Again, the OGC DGGS Abstract Specification enforces no specific implementations of these functional elements, but requires their inclusion (in some form) in any compliant DGGS implementation. This both facilitates flexibility and innovation in the design of individual DGGS implementations and ensures the widest scope for interoperability between compliant DGGS implementations. By focusing on end-point functional requirements and not on the methods by which they are achieved, the OGC DGGS Abstract Specification supports interoperability across multiple social and technical domains. This approach also allows for advancements in the technologies that support these functional elements, without requiring the standard to be constantly re-written.

2.5.3 The Future of the DGGS Standard

In support of the wider adoption and implementation of compliant (and interoperable) DGGS implementations, there are a number of initiatives currently underway within both the OGC and the International Standards Organization (ISO). These initiatives include:

1. The publication of the OGC DGGS Abstract Specification as an ISO standard (ISO 19170). By publishing this standard as an ISO standard, it will be possible to reach a wider community of potential DGGS implementers and thus increase the adoption of DGGS technologies.
2. The establishment of an OGC Registry of compliant DGGS implementations. This will facilitate the certification and publication of compliant DGGS implementations and increase the awareness of the choices of available DGGS implementations that can be applied to a Spatial Data Infrastructure. This will be similar in nature to the Coordinate Reference System Registry. The first release of the OGC DGGS Registry is anticipated to occur by the end of 2018.
3. The development of a standardized specification of a common API language for DGGS. This work is in its early phase but is expected to result in the drafting and publication of a new OGC implementation standard that specifies a common API language supporting and facilitating interoperability between different DGGS implementations. A common API language for DGGS implementations will further lower the technical barriers to the wider implementation of DGGS technologies.

2.5.4 Linkages Between DGGS and Other Standards Activities

As a technology, DGGSs have the potential to impact on almost all spatial technologies and their related standards. Consequently, a number of international standards activities have included references to DGGSs and their potential applications to several scenarios relevant to these initiatives. Two examples of this include:

1. The Joint OGC-W3C Spatial Data on the Web Best Practices (Van Den Brink et al. 2019), where DGGS was proposed as an enabling component of QB4ST (an extension of existing RDF Datacube vocabularies to support spatio-temporal data).
2. The Global Statistical Geospatial Framework (GSGF), adopted during the 6th Session of the United Nations Committee of Experts on Global Geospatial Information Management in August 2016, refers to DGGS and acknowledges that these technologies have the potential to help realize the implementation of the GSGF.

As the number of DGGS implementations increases, so too will the suite of international standards that support them and their applications. The challenge for the International Standards Community will be to keep the number and complexity of these standards to an acceptable level in order to ensure that the DGGS standards do not become a barrier to adoption in themselves.

Acknowledgements Elements of this chapter are inspired by or reproduced from:
• Mahdavi-Amiri, A., Alderson, T., & Samavati, F (2015). A Survey of Digital Earth. *Computers and Graphics, 53 (Part B), 95–117.*
• Foster, C., Purss, M., Peterson, P., Gibb, R., Oliver, S., Samavati, F., Woodcock, R., Evans, B., OGC Discrete Global Grid System (DGGS) Core Standard, 2015.
• Mahdavi-Amiri, A., ACM: Atlas of Connectivity Maps. Ph.D. Thesis, Department of Computer Science, University of Calgary, 2015.

References

Bai J, Zhao X, Chen J (2005) Indexing of the discrete global grid using linear quadtree. In: Jiang J (ed) ISPRS workshop on service and application of spatial data infrastructure, Hangzhou, 14–16 October 2005. ISPRS, pp 267–270

Bartholdi JJ, Goldsman P (2001) Continuous indexing of hierarchical subdivisions of the globe. Int J Geogr Inf Sci 15(6):489–522

Bar-Zeev A (2007) How Google Earth [really] works. Reality Prime. http://www.realityprime.com/blog/2007/07/how-google-earth-really-works. Accessed 30 July 2019.

Ben J, Tong X, Chen R (2010) A spatial indexing method for the hexagon discrete global grid system. In: 18th international conference on geoinformatics, IEEE, Beijing, 18–20 June 2010

Ben J, Li Y, Zhou C et al (2018) Algebraic encoding scheme for aperture 3 hexagonal discrete global grid system. Sci China Earth Sci 61(2):215–227

Bernardin T, Cowgill E, Kreylos O et al (2011) Crusta: a new virtual globe for real-time visualization of sub-meter digital topography at planetary scales. Comput Geosci 37(1):75–85

Cesium Consortium (2019) CesiumJS - Geospatial 3D mapping and virtual globe platform. https://cesiumjs.org. Accessed 30 July 2019

Cozzi P, Ring K (2011) 3D engine design for virtual globes. CRC Press, Hoboken

CSIRO (2019) C-squares home page. http://www.cmar.csiro.au/csquares. Accessed 30 July 2019

Dutton G (1991) Zenithial orthotriangular projection: a useful if unesthetic polyhedral map projection to a peculiar plane. In: 10th international symposium on computer-assisted cartography. Auto-Carto 1991, Baltimore, 25–28 March 1991. ACSM and ASPRS, Bethesda, pp 77–95

Dutton GH (1999) A hierarchical coordinate system for geoprocessing and cartography. Springer, New York

Fan X, Du X, Tan J et al (2009) Three-dimensional visualization simulation assessment system based on multi-source data fusion for the Wenchuan earthquake. J Appl Remote Sens 3(1):1–9

Fekete G, Treinish LA (1990) Sphere quadtrees: a new data structure to support the visualization of spherically distributed data. In: Farrell EJ (ed) Proceedings volume 1259, extracting meaning from complex data: processing, display, interaction, Santa Clara, 11–16 February 1990. SPIE, pp 242–254

Gargantini I (1982) An effective way to represent quadtrees. Commun ACM 25(12):905–910

Global Grid Systems (2019) Global grid systems. https://www.globalgridsystems.com. Accessed 30 July 2019

Goodchild MF (2000) Discrete global grids for digital Earth. In: International conference on discrete global grids, NCGIA, Santa Barbara, 26–28 March 2000

Goodchild MF, Shiren Y (1992) A hierarchical spatial data structure for global geographic information systems. CVGIP: Gr Model Image Process 54(1):31–44

Google Inc (2019a) Google Earth. https://www.google.com/earth. Accessed 30 July 2019

Google Inc (2019b) Tile overlays | maps SDK for android | Google developers. https://developers.google.com/maps/documentation/android-sdk/tileoverlay. Accessed 30 July 2019

Gorski KM, Hivon E, Banday AJ et al (2005) HEALPix: a framework for high-resolution discretization and fast analysis of data distributed on the sphere. Astrophys J 622(2):759–771

Grafarend EW, You RJ, Syffus R (2014) Map projections: cartographic information systems. Springer, Heidelberg
Guo H (2017) Big Earth data: a new frontier in Earth and information sciences. Big Earth Data 1(1–2):4–20
Guo H, Fan X, Wang C (2009) A digital Earth prototype system: DEPS/CAS. Int J Digit Earth 2(1):3–15
Guo HD, Liu Z, Zhu LW (2010) Digital Earth: decadal experiences and some thoughts. Int J Digit Earth 3(1):31–46
Guo H, Liu Z, Jiang H et al (2017) Big Earth Data: a new challenge and opportunity for Digital Earth's development. Int J Digit Earth 10(1): 1–12
Harrison E, Mahdavi-Amiri A, Samavati F (2011) Optimization of inverse snyder polyhedral projection. In: 2011 international conference on cyberworlds, IEEE, Banff, 4–6 October 2011
Harrison E, Mahdavi-Amiri A, Samavati F (2012) Analysis of inverse snyder optimizations. In: Gavrilova ML, Tan CJK (eds) Transactions on computational science XVI. Springer, Heidelberg, pp 134–148
Jet Propulsion Laboratory (2019) Jet propulsion laboratory HEALPix home page. https://healpix.jpl.nasa.gov. Accessed 30 July 2019
Lambers M (2019) ECM: ellipsoidal cube maps. https://marlam.de/ecm. Accessed 30 July 2019
Lambers M, Kolb A (2012) Ellipsoidal cube maps for accurate rendering of planetary-scale terrain data. In: Bregler C, Sander P, Wimmer M (eds) 20th Pacific conference on computer graphics and applications (short papers). Pacific Graphics 2012, Hong Kong, 12–14 September 2012. The Eurographics Association, pp 5–10
Mahdavi-Amiri A, Samavati F (2012) Connectivity maps for subdivision surfaces. In: Richard P, Kraus M, Laramee R, Braz J (eds) International conference on computer graphics theory and applications. GRAPP 2012, Rome, 24–26 February 2012. SciTePress, pp 26–37
Mahdavi-Amiri A, Samavati F (2014) Atlas of connectivity maps. Comput Gr 39:1–11
Mahdavi-Amiri A, Bhojani F, Samavati F (2013) One-to-two digital Earth. In: Bebis G, Boyle R, Parvin B et al (eds) International symposium on visual computing. SVC 2013: advances in visual computing. Springer, Heidelberg, pp 681–692
Mahdavi-Amiri A, Harrison E, Samavati F (2015a) Hexagonal connectivity maps for digital Earth. Int J Digit Earth 8(9):750–769
Mahdavi-Amiri A, Samavati F, Peterson P (2015b) Categorization and conversions for indexing methods of discrete global grid systems. ISPRS Int J Geo-Inf 4(1):320–336
Middleton L, Sivaswamy J (2005) Hexagonal image processing: a practical approach. Springer, London
Mocnik F-B (2019) GitHub - GIScience/geogrid: library for discrete global grid systems. https://github.com/GIScience/geogrid. Accessed 30 July 2019
NASA (2019) NASA WorldWind. https://worldwind.arc.nasa.gov/. Accessed 30 July 2019
Percivall G (2013) Geodata fusion study by the open geospatial consortium. In: Pellechia MF, Sorensen RJ, Palaniappan K (eds) Proceedings volume 8747, geospatial InfoFusion III, Baltimore, 29 April–3 May 2013. SPIE, pp 87470A-1–87470A-13
Peterson P (2004) Close-packed, uniformly adjacent, multiresolutional, overlapping spatial data ordering. US Patent 8400451, 19 March 2013
Presagis (2019) OGC CDB - Presagis. https://www.presagis.com/en/glossary/detail/ogc-cdb. Accessed 30 July 2019
Purss M, Gibb R, Samavati F et al (2017) Discrete global grid systems abstract specification – Topic 21. In: Purss M (ed). Open geospatial consortium, Wayland
Purss M, Peterson P, Strobl P et al (2019) Datacubes: a discrete global grid systems perspective. Cartographica 54(1):63–71
PYXIS Innovation (2011) How PYXIS works - Pyxis public wiki. http://www.pyxisinnovation.com/pyxwiki/index.php?title=How_PYXIS_Works. Accessed 30 July 2019
Rees T (2003) "C-squares", a new spatial indexing system and its applicability to the description

of oceanographic datasets. Oceanography 16(1):11–19

Roşca D, Plonka G (2011) Uniform spherical grids via equal area projection from the cube to the sphere. J Comput Appl Math 236(6):1033–1041

Roşca D, Plonka G (2012) An area preserving projection from the regular octahedron to the sphere. Results Math 62(3):429–444

Sahr K (2008) Location coding on icosahedral aperture 3 hexagon discrete global grids. Comput Environ Urban Syst 32(3):174–187

Sahr K (2016) Central place indexing systems. US Patent 9311350, 12 April 2016

Sahr K, White D, Kimerling AJ (2003) Geodesic discrete global grid systems. Cartogr Geogr Inf Sci 30(2):121–134

Schwartz J (2018) Bing maps tile system. Microsoft Docs. https://docs.microsoft.com/en-us/bingmaps/articles/bing-maps-tile-system. Accessed 30 July 2019

SEEGrid (2019) WebHome<SCENZGrid<SEEGrid. https://www.seegrid.csiro.au/wiki/SCENZGrid/WebHome. Accessed 30 July 2019

Skyline Software Systems (2019) Skyline software systems: skylineglobe 3D earth software solutions. http://www.skylineglobe.com. Accessed 30 July 2019

Snyder J (1992) An equal-area map projection for polyhedral globes. Cartographica 29(1):10–21

Tanner C, Migdal C, Jones M (1998) The clipmap: a virtual mipmap. In: Cunningham S, Bransford W, Cohen MF et al. (eds) Proceedings of the 25th annual conference on computer graphics and interactive techniques. ACM, Orlando, pp 19–24

Telespazio (2019) GlobWeb. https://github.com/TPZF/GlobWeb. Accessed 30 July 2019

Tobler W, Chen Z-T (1986) A quadtree for global information storage. Geogr Anal 18(4):360–371

Tong X, Ben J, Wang Y et al (2013) Efficient encoding and spatial operation scheme for aperture 4 hexagonal discrete global grid system. Int J Geogr Inf Sci 27(5):898–921

van den Brink L, Barnaghi P, Tandy J et al (2019) Best practices for publishing, retrieving, and using spatial data on the web. Semant Web 10(1):99–114

Vince A (2006) Indexing the aperture 3 hexagonal discrete global grid. J Vis Commun Image Represent 17(6):1227–1236

Vince A, Zheng X (2009) Arithmetic and Fourier transform for the PYXIS multi-resolution digital Earth model. Int J Digit Earth 2(1):59–79

White D (2000) Global grids from recursive diamond subdivisions of the surface of an octahedron or icosahedron. Environ Monit Assess 64(1):93–103

White D, Kimerling JA, Overton SW (1992) Cartographic and geometric components of a global sampling design for environmental monitoring. Cartogr Geogr Inf Syst 19(1):5–22

Yu J-Q, Wu L-X (2009) Spatial subdivision and coding of a global three-dimensional grid: Spheoid degenerated-octree grid. In: 2009 IEEE international geoscience and remote sensing symposium, IEEE, Cape Town, 12–17 July 2009

Yu JQ, Wu LX, Zi GJ et al (2012) SDOG-based multi-scale 3D modeling and visualization on global lithosphere. Sci China Earth Sci 55(6):1012–1020

Troy Alderson is a Ph.D. candidate in the Department of Computer Science at the University of Calgary, where he also received his M.Sc. in Computer Science (2014) and B.Sc.'s in Computer Science and General Mathematics (2010). His current work focuses on developing unified frameworks for subdivision and multiresolution for various types of geometric shapes.

Matthew Purss is a founding co-chair of the Open Geospatial Consortium's Discrete Global Grid Systems working groups. He holds a Ph.D. in Exploration Geophysics and throughout his 20+ year career he has been a leader in the theoretical development, practical implementation and standardisation of advanced spatial data infrastructures and grid systems.

Xiaoping Du is currently an Associate Professor with the Institute of Remote Sensing and Digital Earth, Chinese Academy of Sciences. His research interests include Digital Earth and geospatial big data analytics, coastal flooding risk assessment due to global climate change, and remote sensing for natural disasters & environmental monitoring.

Ali Mahdavi-Amiri is a University Research Associate at Computing Science department in Simon Fraser University. His research interests is in Computer Graphics with specialty in Geometric Modeling, Geospatial Modeling and Visualization, Digital Fabrication, and Machine Learning. Ali has received several awards including NSERC Postdoctoral Fellowship, Graduate Achievement Award, Michael A. J. Sweeney Award, and Teaching Excellence Award.

Faramarz Samavati is a professor of Computer Science at the University of Calgary. Dr. Samavati's research interests include Computer Graphics, Visualization, and Digital Earth. Over the past eight years, he has received seven best paper awards, Digital Alberta Award, Great Supervisor Award, and University of Calgary Peak Award.

3

Remote Sensing Satellites: Developments and Trends

Wenxue Fu, Jianwen Ma, Pei Chen and Fang Chen

Abstract The term remote sensing became common after 1962 and generally refers to nonintrusive Earth observation using electromagnetic waves from a platform some distance away from the object of the study. After more than five decades of development, humankind can now use different types of optical and microwave sensors to obtain large datasets with high precision and high resolution for the atmosphere, ocean, and land. The frequency of data acquisition ranges from once per month to once per minute, the spatial resolution ranges from kilometer to centimeter scales, and the electromagnetic spectrum covers wavebands ranging from visible light to microwave wavelengths. Technological progress in remote sensing sensors enables us to obtain data on the global scale, remarkably expanding humanity's understanding of its own living environment from spatial and temporal perspectives, and provides an increasing number of data resources for Digital Earth. This chapter introduces the developments and trends in remote sensing satellites around the world.

Keywords Remote sensing · Digital Earth · Satellite · Earth observation

3.1 Development of Remote Sensing

Remote sensing is a core technology for Earth observation. It covers information collection, in-orbit processing, information storage and transmission, ground reception, processing for applications, calibration, verification, applied research, and basic research, providing fundamental data resources for Digital Earth (Guo 2012).

W. Fu (✉) · J. Ma · F. Chen
Institute of Remote Sensing and Digital Earth, CAS, Beijing, China
e-mail: fuwx@aircas.ac.cn

P. Chen
School of Astronautics, Beihang University, Beijing, China

3.1.1 Overview of Remote Sensing

3.1.1.1 Remote Sensing Platforms

Remote sensing refers to various observation and exploration activities of the environment involving humans and photoelectronic devices carried by satellites, spacecraft (including space shuttles), aircraft, near-space vehicles, and various terrestrial platforms. Artificial satellites that carry sensors to capture images of Earth's surface are referred to as remote sensing satellites. Satellites can successively observe the whole globe or an assigned part of it within a defined time period (Guo et al. 2016). Aircraft often have a definite advantage because of their mobilization flexibility. They can be deployed wherever and whenever weather conditions are favorable. Satellites and aircraft collect the majority of base map data and imagery used in remote sensing, and the sensors typically deployed on these platforms include film and digital cameras, light-detection and ranging (LiDAR) systems, synthetic aperture radar (SAR) systems, and multispectral and hyperspectral scanners. Many of these instruments can also be mounted on land-based platforms such as vans, trucks, tractors, and tanks. In the future, the Moon will also be an ideal remote sensing platform (Guo et al. 2014a, 2018).

3.1.1.2 Remote Sensing Sensors

There are several types of Earth observation sensors: photographic sensors, scanning imaging sensors, radar imaging sensors, and nonimaging sensors. Photographic sensors work like a digital camera. Scanning imaging sensors capture two-dimensional images by scanning point by point and line by line in a time sequence. These are widely used today; such sensors can be further divided into surface scanning and image scanning sensors. Imaging radar is an active sensor that emits electromagnetic waves to form a lateral profile. Currently, most Earth observation satellites carry SAR systems that feature very high resolutions.

In the early stage of spaceborne Earth observation, traditional film-based imaging devices, return beam vidicon (RBV) TV cameras, and optical scanners were the main devices used for Earth observation. Images obtained from these devices were mainly color and black-and-white representations of Earth's surface and cloud layer, covering the visible light and near infrared ranges. After the first land observation satellite, Landsat 1, was launched in 1972, the new multispectral scanner (MSS) it carried sent data that was processed in the form of a digital time sequence array. This marked a progressive step in the development of digital image processing.

Compared with optical remote sensors, SARs work in various weather conditions and can penetrate some surface objects. In contrast to passive sensor systems that only receive reflected solar light or infrared radiation, radar systems act as active sensors and emit electromagnetic waves on their own. A radar sensor sends pulses of energy to the Earth's surface and part of that energy is reflected and forms return signals. The strength of the return signal depends on the roughness and dampness of the Earth's surface and the inclination of surface objects toward the waves sent by radar.

3.1.2 Development of Remote Sensing Satellites

Based on a life cycle of approximately thirteen years, Earth observation satellites have gone through four generations (Fig. 3.1) (Zhou 2010).

(1) The first generation, beginning spaceborne Earth observation: 1960–1972

CORONA, ARGON, and LANYARD were the first three imaging satellite observation systems. Data obtained from these satellites were used for detailed terrestrial reconnaissance and regional mapping. In the early years, satellite images were made by combining hundreds or even thousands of photos, most of which were black-and-white, with a small number of color photos or three-dimensional image pairs. These images covered most parts of Earth. For example, images obtained using the KH-5 camera covered most of the Earth's surface with a 140-m pixel resolution. However, these images did not form systematic observations like those achieved later with Landsat data.

(2) The second generation, experimental and tentative application: 1972–1986

Landsat-1 was launched on July 23, 1972, marking the start of modern satellite-carried Earth observation. It provided a novel high-resolution Earth image database to international science organizations, making further exploration of Earth's resources possible. Landsat-1 carried an MSS that received four bands with wavelengths from 0.5 to 1.1 μm with a spatial resolution of 80 m, frame width of 185 km, and revisit cycle of eighteen days. Notably, Landsat-1 transmitted data in digital form for the first time. The foundation for multispectral processing was laid in the 1970s and organizations involved in this field included NASA, Jet Propulsion Laboratory (JPL), United States Geological Survey (USGS), Environmental Research Institute of Michigan (ERIM), and Laboratory for Applications of Remote Sensing (LARS). Ten years

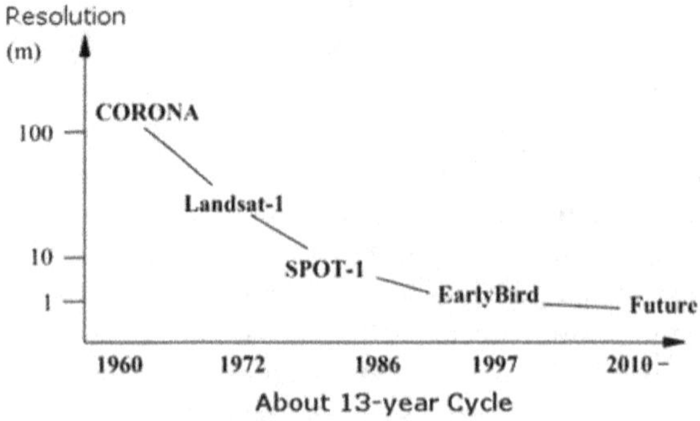

Fig. 3.1 History of the thirteen-year cycle of Earth observation satellite development (Zhou 2010)

later, Landsat accommodated four more MSS wavebands as Landsat TM emerged during 1982–1984 with a spatial resolution of 30 m covering seven spectral bands. Soon afterwards, the famous SPOT HRV system was launched in 1986 with a spatial resolution of 10 m for the panchromatic wavebands and 30 m for three other multispectral bands.

(3) The third generation, wide application: 1986–1997

After 1986, the technology and applications of satellite Earth observation developed rapidly. SPOT-1, launched on February 22, 1986, carried a high-resolution visual sensor and was the first use of pushbroom linear array sensors. It was also the first satellite system capable of cross-track three-dimensional observation. Later, the ESA launched the ERS-1 SAR on July 17, 1991. ERS-1 was an active microwave satellite that provided images with a spatial resolution of 30 m. Japan launched its JERS-1 in February 1992 with an L-band SAR, building up the overall observation capacity of SARs. Data provided by these active microwave sensors played an important role in enhancing the observation and understanding of environmental and climatic phenomena, and supported the categorization of sea ice and research on the coastal zone.

(4) The fourth generation, high-resolution and hyperspectral imaging: 1997–2010
This comprises the latest generation of Earth observation satellites equipped with the most advanced technologies that are still gradually maturing. The main features are a spatial resolution of 1 m or less, coverage of 200 wavebands ranging from 0.4 to 2.5 μm in wavelength, a spectral resolution of 10 nm, revisit cycles less than three days, capability of multiangle and three-dimensional observation, and precise spatial positioning with GPS. The major advantage of high-resolution imaging is that it allows for identification of buildings, roads, and modern construction projects as well as change detection. As a result, high-resolution imagery products are mainly used in GIS and special-purpose mapping.

At this stage, attention was primarily focused on spatial and temporal resolutions, spectral coverage, orbital height, revisit capability, mapping bandwidth, image dimensions, capacity for three-dimensional observation, imaging models, data storage, and the market demand for satellites.

(5) The fifth generation, a new era of satellite Earth observation

Next-generation Earth observation satellites are expected to be highly intelligent and integrate Earth observation sensors, data processing devices, and communication systems. Global surveying and real-time environmental analysis of Earth will become possible. More experts as well as casual users will be involved in remote sensing, photogrammetry and GIS, and data inversion products will also be updated more frequently. To achieve real-time data acquisition, improve applications and spare casual users the trouble of understanding complicated data processing, image providers will offer mature imaging products that directly meet various demands (Guo et al. 2014b).

3.2 Land Observation Satellites

Land observation satellites have been developed for land resource investigation, terrestrial environment research, crop condition forecasting, and natural disaster monitoring. Terrestrial variables have a specific “ground object spectrum” and radiation scattering; terrestrial variables can be retrieved by considering the direction, scale, and sensitivity to establish the relationship between electromagnetic waves and ground surface variables for space observation.

3.2.1 US Land Observation Satellites

The United States launched its first land satellite, Landsat 1, on July 23, 1972. For the first time in human history, satellites were consistently providing Earth images with a certain resolution, making it possible to use satellites to survey Earth’s resources. Since then, the country has launched seven satellites in the Landsat series (the launch of Landsat 6 failed). They are currently the world’s most widely used land observation satellites (Table 3.1).

Later, the United States launched a series of high-resolution commercial remote sensing satellites. The IKONOS satellite, launched on September 24, 1999, was the world’s first commercial remote sensing satellite providing high-resolution images. After that, the country launched the QuickBird, WorldView-1, GeoEye-1, and Wor1dView-2 satellites in October 2001, September 2007, September 2008, and October 2009, respectively, with improved resolutions from 0.61 to 0.41 m (multispectral) (Aguilar et al. 2013).

3.2.1.1 Landsat Program

The Earth Resources Satellite Program involves a series of Earth observation satellites jointly managed by NASA and the United States Geological Survey (USGS). These satellites collect information about Earth from space. They have been providing digital photos of Earth’s continents and coastal regions for more than 40 years, enabling researchers to study Earth from various aspects and evaluate the impacts of natural and human activities on the dynamics of the Earth system.

(1) Landsat 7

Landsat 7 moves around Earth on a near-polar sun-synchronous orbit, with an orbital altitude of 705.3 km and an operation cycle of 98.9 min, covering Earth once every sixteen days. During the day, it operates on a descending orbit, crossing the equator at 10:00 AM. The orbit is adjusted so that orbital inclination is kept within a certain limit and the deviation of the satellite transit time from the nominal time is kept within ±5 min.

Table 3.1 Land satellites launched by the United States

Satellite code	Type of orbit	Orbital altitude (km)	Orbital period (min)	Orbital inclination (°)	Launch date
Landsat-1	Sun-synchronous orbit	917	103.1	99.2	1972.6.23
Landsat-2	Sun-synchronous orbit	917	103.3	99.2	1975.1.22
Landsat-3	Sun-synchronous orbit	917	103.1	99.1	1978.3.5
Landsat-4	Sun-synchronous orbit	705	98.9	98.2	1982.7.16
Landsat-5	Sun-synchronous orbit	705	98.9	98.2	1984.3.1
TRMM	Inclined orbit	405	93.5	35	1997.11.27
Landsat-7	Sun-synchronous orbit	705	98.9	98.2	1999.4.15
Terra	Sun-synchronous orbit	705	99	98.2	1999.12.18
ACRIMSAT	Sun-synchronous orbit	716	90	98.13	1999.12.20
GRACE	Polar orbit	400	94	89	2002.3.17
Aqua	Sun-synchronous orbit	705	98.8	98.2	2002.5.4
ICESat	Inclined orbit	600	97	94	2003.1.12
SORCE	Inclined orbit	600	90	40	2003.1.25
Suomi NPP	Sun-synchronous orbit	824	101	98.7	2011.10.28
Landsat-8	Sun-synchronous orbit	705	99	98.2	2013.2.12

Table 3.2 ETM+ bands

Waveband	Wavelength range (μm)	Ground resolution (km)
1	0.45–0.515	30
2	0.525–0.605	30
3	0.63–0.690	30
4	0.75–0.90	30
5	1.55–1.75	30
6	10.40–12.50	60
7	2.09–2.35	30
Pan	0.52–0.90	15

The ETM+ of Landsat 7 was developed based on the TM of Landsats 4 and 5 and the ETM of Landsat 6. It is a multispectral vertical-orbit scanning radiometer that performs Earth imaging directly facing the nadir and obtains high-resolution ground images. Its scanning width is 185 km. Similar to the previous Landsats, the EMT+ uses a scan line corrector to eliminate the interline overlap or interline spacing caused by the scanning operation or orbital motion.

In the visible and near-infrared (VNIR) range, ETM+ has four color bands and one panchromatic band. Each of the six sounder arrays in the visible, near-infrared and SWIR bands has sixteen sounders staggered along the orbital direction, and each sounder corresponds to a ground area of 30 × 30 m. The LWIR sounder array has eight sounders, each corresponding to a ground area of 60 × 60 m, with a resolution twice as high as that of the previous thermal infrared TM. The panchromatic band was a new addition to Landsat 7. The sounder array consists of 32 sounders, each corresponding to a ground area of 15 × 15 m. The bands of ETM+ are described in Table 3.2.

(2) Landsat 8 (LDCM)

Landsat 8, also referred to as LDCM, carries two main payloads: one operational land imager (OLI) and one thermal infrared sensor (TIRS). Compared with the payloads of previous Landsats, the performance of the OLI and TIRS are much improved.

Landsat 8 can capture at least 400 images per day (its predecessors could only capture 250). This is because Landsat 8 is more flexible in monitoring an area (Ali et al. 2017). Previous Landsats could only monitor a certain swath of land directly under their flight path, but the remote sensor of Landsat 8 can capture information about land that deviates from the flight path by a certain angle, which the previous Landsats could do only in subsequent laps. This advantage helps capture imagery needed for multitemporal comparison (such as images concerning disasters).

The main parameters of Landsat 8 are: a Worldwide Reference System-2 (WRS-2) flight path/line system, a sun-synchronous orbital altitude of 705 km, global coverage cycle of sixteen days (except for high-latitude polar regions), 233 orbits per cycle, an orbital inclination of 98.2° (slightly to the right), an operation cycle of 98.9 min, and a 170 × 185 km imaging area. The satellite crosses the equator at 10:00 AM ±

15 min. Its image directory is prepared in the same way as those of Landsats 4, 5 and 7, and it supports the ability to capture the main image and images that deviate from the nadir point to a limited extent (±1 flight path/line).

3.2.1.2 GRACE Satellite Program

The Gravity Recovery and Climate Experiment (GRACE) satellite program aims to obtain the features of medium and long waves of Earth's gravity field and the time-varying characteristics of the global gravity field (Melzer and Subrahmanyam 2017) and to sound the atmospheric and ionospheric environment. The GRACE satellite was launched on March 17, 2002 from the Plesetsk Launch Center in northern Russia. Its working principle is shown in Fig. 3.2.

The satellite adopts a low-low satellite-to-satellite tracking mode with two simultaneously launched low Earth orbit satellites that travel on the same orbit with a distance of 220 km in between them. Satellite-borne GPS receivers can accurately determine the orbital position of the two satellites and measure their distance and the changes in distance accurate to the micron level. A triaxial accelerometer is used to measure nonconservative forces. The observation data of each satellite, including the data of gravity-related measurements and GPS occultation measurements, are transmitted to the ground station via S-band radio waves.

The scientific objectives of the GRACE satellite project are (1) to determine Earth's mediumwave and longwave gravity field with a geoid precision of 0.01 cm and 0.01 mm for 5,000 km and 500 km wavelengths, respectively, which is two orders of magnitude higher than that of the CHAMP satellite (Ditmar 2004); (2) to

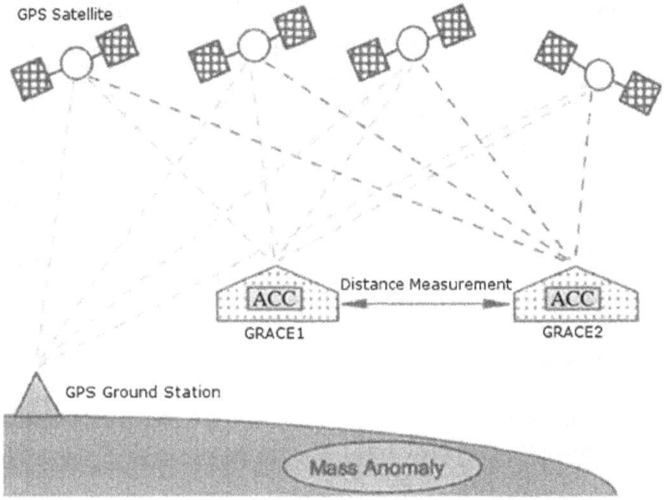

Fig. 3.2 GRACE working principle (Lu 2005)

determine changes in the global gravity field based on observation data from 2 to 4 weeks or longer, with an expected geoid determination precision of 0.001 mm/y; and (3) to sound the atmospheric and ionospheric environment. As the GRACE satellites provide highly accurate information about Earth's mediumwave and longwave gravity field and its time-dependent changes, they mark the beginning of a new era of satellite-based gravity research (Liu 2009).

3.2.1.3 Commercial Remote Sensing Satellites

On September 24, 1999, the IKONOS satellite was successfully launched at Vandenberg Air Force Base, marking the start of the era of high-resolution commercial satellites. On March 31, 2015, IKONOS was retired after 15 years of over service, a working lifetime more than twice of that in the design. IKONOS was a commercial satellite that acquired 1-m resolution panchromatic images and 4-m resolution multispectral images. Additionally, the resolution of the integrated color image with the panchromatic and multispectral images was up to 1 m. The IKONOS revisit period was 1–3 days imaging from the 681 km orbit.

The QuickBird satellite was launched in October 2001 with a panchromatic spatial resolution of 0.61 m and multispectral resolution of 2.44 m. The WorldView-1 satellite, launched on September 18, 2007, was the commercial imaging satellite with the highest resolution and the fastest response speed in the world at that time. WorldView-1 has an average revisit period of 1.7 days in a sun-synchronous orbit at an altitude of 496 km and inclination angle of 98°. The large-capacity panchromatic system can capture images up to 550,000 km^2 with 0.5-m resolution every day. The satellite also has high geolocation accuracy and quick response, which provides quick aiming at the target to effectively perform on-track stereo imaging. Its acquisition capacity is four times that of the QuickBird satellite. Parameters of the WorldView-1 satellite are shown in Table 3.3.

WorldView-2, launched in October 2009, was the first commercial remote sensing satellite in the world to provide 8-band high resolution data, greatly enhancing the customer service ability of DigitalGlobe. In June 2014, with the consent of the US Department of Defense and the State Department, the US Department of Commerce formally approved DigitalGlobe's application for the sale of 0.25-m resolution satellite image data.

With the implementation of the new policy, WorldView-3, the third-generation remote sensing satellite, was successfully launched in August 2014 and is the world's first commercial multipayload, hyperspectral and high resolution satellite, providing 0.31-m panchromatic imagery and 1.24-m multispectral imagery. The WorldView-4 commercial remote sensing satellite was launched in November 2016 and has greatly improved the overall data acquisition capability of the DigitalGlobe constellation group. It can image any point on the Earth 4.5 times a day, with a ground sampling distance (GSD) of less than 1 m.

Table 3.3 WorldView-1 satellite parameters

Parameter	Value
Orbit	Solar synchronization at a height of 450 km
Satellite size, weight and power supply	3.6 m high, 2.5 m wide; the total span of the solar panels is 7.1 m; weight of 2500 kg; 3.2 kw solar cells
Remote sensor band	Panchromatic
Resolution	Subsatellite point: 0.45 m (GSD)
Swath	Subsatellite point: 16 km
Altitude measurement and control	Tri-axial stability
Data transmission	Image and auxiliary data: 800 Mbit/s, X-band
Data acquisition for each orbit	331 Gbit
Maximum continuous imaging area of a single-circle orbit	60 × 60 km (equivalent to 4 × 4 square images); 30 × 30 km (equivalent to 2 × 2 square images)
Revisit period	While imaging with 1 m GSD: 1.7 days

3.2.1.4 Satellite Images for Google Earth

Google Earth's images come from multisource data composed of satellite images and aerial data. Its satellite images mainly come from the QuickBird commercial satellite, GeoEye satellite and IKONOS satellite of the DigitalGlobe Company of the United States, as well as the SPOT-5 satellite of France.

The GeoEye series of satellites are the next generation of the IKONOS and OrbView satellites. The GeoEye-1 satellite, launched on September 6, 2008 from Vandenberg Air Force Base in California can acquire black-and-white (panchromatic) imagery with 0.41-m resolution and color (multispectral) imagery with 1.65-m resolution, and can accurately locate the target position with 3 m accuracy. Therefore, it has become the most powerful commercial imaging satellite with the highest resolution and accuracy in the world. The GeoEye-1 satellite runs in a solar synchronous orbit with an altitude of 681 km and inclination angle of 98°, an orbit period of 98 min and a revisit period of less than 3 days. The satellite's launch mass was 1955 kg, and the design life is 7 years. The payload of the GeoEye-1 satellite is a pushbroom imaging camera consisting of an optical subsystem (telescope module, aperture 1.1 m), a focal plane module and a digital electronic circuit. The main parameters of the GeoEye-1 satellite are shown in Table 3.4.

Table 3.4 The main parameters of the GeoEye-1 satellite

Parameter		Values
Resolution		Subsatellite point panchromatic: 0.41 m, side-looking 28° panchromatic: 0.5 m, subsatellite point multispectral: 1.65 m
Swath		Subsatellite point: 15.2 km; single scene 225 km^2 (15 × 15 km)
Camera mode		Panchromatic and multispectral simultaneous (panchromatic fusion), monochromatic and monochromatic
Revisit period		2–3 days
Wavelength	Panchromatic	450–800 nm
	Multispectral	Blue: 450–510 nm
		Green: 510–580 nm
		Red: 655–690 nm
		Near-infrared: 780–920 nm

3.2.2 European Land Observation Satellites

3.2.2.1 ESA Satellites

(1) *CryoSat-2*

On April 8, 2010, the ESA launched CryoSat-2 using a Dnepr rocket. As one of the primary missions of the European Earth Observation Program (EOP), CryoSat uses a radar altimeter to measure the thickness of Earth's land ice and sea ice sheets, especially polar ice and oceanic floating ice, to study the effects of global warming. Earlier, in October 2005, the launch of CryoSat-1 was unsuccessful due to a rocket failure.

SIRAL is the main payload of CryoSat-2, weighing 62 kg (Dibarboure et al. 2011). It is mainly used to observe the internal structure of ice shields and study sea ice and landforms. SIRAL has three measurement modes: the low-resolution measurement (LRM) mode, which is only used to measure relatively flat polar and oceanic ice sheets; the SAR mode that is used to measure sea ice with an along-track resolution of 250 m; and the InSAR mode that is used to study ice sheets in more complex and steep areas with a measurement accuracy of 1 to 3 cm (Wingham et al. 2006). In contrast to traditional radar altimeters, the delay Doppler radar altimeter (DDA) adopted by SIRAL can emit continuous pulse trains and can make efficient use of Earth's surface reflection power via full Doppler bandwidth. SIRAL was designed based on existing instruments but has improved performance compared with the radar altimeters on board ERS-1, ERS-2 and ENVISAT. SIRAL has two pairs of Cassegrain antennas that are used to transmit radar signals and receive signals reflected from the ground to obtain accurate information about polar and sea ice thickness. SIRAL can accurately measure irregular and steep edges of land ice, and can obtain data from sea and river ice. The characteristics of SIRAL are shown in Table 3.5.

Table 3.5 SIRAL characteristics

Parameter	Mode of measurement		
	LRM	SAR	InSAR
Receiving chain	1 (left)	1 (left)	2 (left and right)
Sampling interval (m)	0.47	0.47	0.47
Bandwidth (MHz)	350	350	350
Pulse repetition frequency (PRF) (Hz)	1,970	17.8	17.8
Transmitter pulse width (μs)	49	49	49
Effective echo width (μs)	44.8	44.8	44.8
Pulse duration (ms)	None	3.6	3.6
Color synchronization pulse	None	64	64
Color synchronization pulse period (ms)	None	11.7	46.7
Tracking pulse bandwidth (MHz)	350	350	40
Average tracking pulse/46.7 ms	92	32	24
Data transmission rate (Mbps)	0.051	11.3	11.3 (2)
Power consumption (W)	95.5	127.5	127.5

(2) *Copernicus Program*

The Copernicus program, formerly Global Monitoring for Environment and Security (GMES), was a major space development program launched by the European Union in 2003. Its main purpose is to ensure Europe's sustainable development, enhance international competitiveness, security and to realize real-time dynamic monitoring of the environment by coordinating, managing and integrating the observation data of existing and future European and non-European (third-party) satellites.

In terms of EOS infrastructure development, the GMES program is divided into three parts. The first part is the space-based observation for which ESA is responsible. New satellites will be launched and the existing satellites are divided into six mission groups (see Table 3.6). The second part is the ground-based observation for which the European Environment Agency (EEA) is responsible. The third part is data sharing, which calls for building capacity for comprehensive and sustainable observation data applications and the construction of network entrances for data access; data services are mainly provided by the ESA, French Space Agency (CNES), and EUMETSAT.

3.2.2.2 France's Satellites

On February 22, 1986, France launched its first Earth resources observation satellite, SPOT-1. Thus far, seven SPOT satellites have been sent into space. The sounders adopted by these satellites have unique characteristics and the imaging method is also unique. Additionally, SPOT satellites are the world's first remote sensing satellites

Table 3.6 The Copernicus (GMES) space segment

Satellite	Function	Purpose	Launch date
Sentinel 1	SAR imaging	Continuous all-weather monitoring of ships and oil spills, other applications	Sentinel 1A: 2014.4.3 Sentinel 1B: 2016.4.25
Sentinel 2	Multispectral imaging	Land applications such as for cities, forests, agriculture, etc.	Sentinel 2A: 2015.6.23 Sentinel 2B: 2017.3.7
Sentinel 3	Ocean and land monitoring	Ocean color, vegetation, sea surface and land surface temperatures, sea wave height, etc.	Sentinel 3A: 2016.2.16 Sentinel 3B: 2018.4.25
Sentinel 4	Geosynchronous orbit—atmospheric monitoring	Monitoring of atmospheric composition and boundary layer pollution	
Sentinel 5	Low-orbit atmospheric research satellite	Monitoring of atmospheric composition	Sentinel 5P: 2017.10.13
Sentinel 6	Non-sun-synchronous orbit at 1,336 km mean altitude	Providing reference continuity and a high-precision ocean topography service after Jason-3	

to have stereo imaging capability. Basic information on the SPOT series is shown in Table 3.7.

The CNES launched the SPOT-5 remote sensing satellite in May 2002, with a design life of five years and total mass of 3,030 kg. Compared with the first four SPOT satellites, SPOT-5 has significantly improved observation capability and incorporated new instruments (Table 3.8), including the following: (1) An HSR with a panchromatic spectral resolution of 10 m, (2) two HRGs with working bands that differ from HRV and HRVIR, and (3) a VEGETATION-2 imager that could achieve global coverage almost every day with an imaging resolution of 1 km.

SPOT-6 was launched by India's Polar Satellite Launch Vehicle on flight C21 on September 9, 2012 and SPOT-7 was launched on PSLV flight C23 on June 30, 2014. They form a constellation of Earth-imaging satellites designed to provide continuity of high-resolution, wide-swath data up to 2024. EADS Astrium took the decision to build this constellation in 2009 based on a perceived government need for this kind of data. SPOT-6 and SPOT-7 are phased in the same orbit as Pléiades 1A and Pléiades 1B, which are at an altitude of 694 km, forming a constellation of 2-by-2 satellites that are 90° apart from one another.

Table 3.7 SPOT satellite information

Satellite	Launch date	Sensor	Service period (year)	Width (km)	Altitude (km)
SPOT-1	1986.02.22	Stereo imaging system with a pushbroom scanner (HRV)	1986–1990	2 × 16	830
SPOT-2	1990.01.22	Stereo imaging system with a pushbroom scanner (HRV)	1990–2006	2 × 16	830
SPOT-3	1993.09.26	Improved HRV, solid altimeter, laser reflector	1993–1996	110–2,000	832
SPOT-4	1998.03.24	Improved HRV, HRVIR	1998–2013	110–2,200	1,334
SPOT-5	2002.05.03	HRG, HRVIR, HSR	Still in operation	60 × 60–60 × 120	830
SPOT-6	2012.09.09	Multispectral Imagery	Still in operation	60 × 60	695
SPOT-7	2014.06.30	Multispectral Imagery	Still in operation	60 × 60	695

Table 3.8 Technical parameters of the three sensors on board SPOT-5

Type of remote sensor	Waveband	Wavelength range (μm)	Resolution (m)	Width (km)
HRG	Panchromatic	0.49–0.69	2.5 or 5	60
HRVIR	Multispectral	0.49–0.61	10	60
		0.61–0.68	10	60
		0.78–0.89	10	60
		1.58–1.75	20	60
		0.43–0.47	1,000	2,250
		0.61–0.68	1,000	2,250
		0.78–0.89	1,000	2,250
		1.58–1.75	1,000	2,250
HSR	Panchromatic	0.49–0.69	10	120

3.2.2.3 Germany's Satellites

CHAMP is a small satellite mission for geoscience research, atmospheric studies, and applications headed by the German Research Centre for Geosciences (GFZ) (GFZ 2018; Guo et al. 2008). As a near-polar, low Earth orbit satellite equipped with high-precision, multifunction, completely satellite-borne instruments (magnetometer, accelerometer, STAR sensor, GPS receiver, laser mirror, ion drift meter). CHAMP had a design life of five years, and ended on September 19, 2010. Its shape and onboard instruments are shown in Fig. 3.3. It could simultaneously measure Earth's gravitational and magnetic fields with high precision and detect their temporal and spatial changes (Baduraet al. 2006).

The CHAMP mission had three main goals: (1) to accurately determine the long-wavelength characteristics of the Earth's gravitational field and its temporal changes; (2) to estimate, with unprecedented accuracy, temporal and spatial variations of the magnetic field of the Earth's main body and crust, and all components of the magnetic field; and (3) to study temperature, water vapor, and electrons using a large amount of globally distributed GPS signal refraction data generated by the atmosphere and ionosphere.

TerraSAR-X is a German SAR satellite mission for scientific and commercial applications that was launched on June 15, 2007. The project is managed by the DLR (German Aerospace Center). In 2002, EADS Astrium GmbH was awarded a contract to implement the X-band TerraSAR satellite (TerraSAR-X) on the basis of a public-private partnership agreement (PPP). In this arrangement, EADS Astrium funded part of the implementation cost of the TerraSAR-X system.

The science objectives are to make multimode and high-resolution X-band data available for a wide spectrum of scientific applications in fields such as hydrology, geology, climatology, oceanography, environmental and disaster monitoring, and cartography (DEM generation) using interferometry and stereometry.

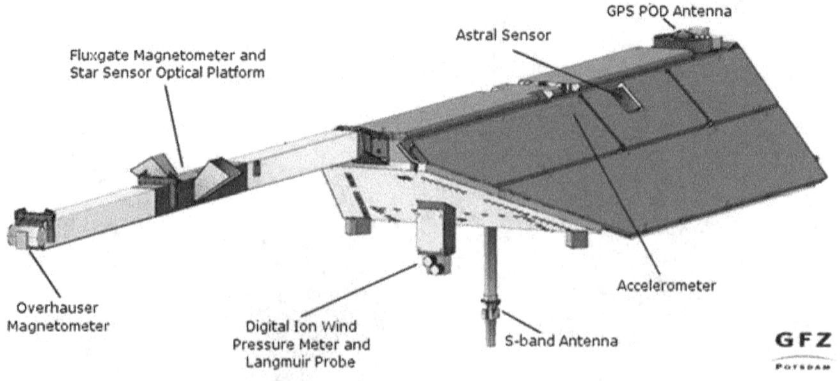

Fig. 3.3 CHAMP satellite structure (GFZ 2018)

3.2.3 *China's Land Observation Satellites*

3.2.3.1 Resource Satellites

Resource satellites are used to survey the Earth's natural resources and carry out scientific research on the Earth system. China has developed a series of satellites for land observation.

(1) *CBERS satellites*

The China-Brazil Earth Resource Satellites (CBERS) were jointly developed by China and Brazil using their combined investment in accordance with an agreement signed by both countries in 1988. CBERS was shared by the two countries after being put into operation. The first CBERS (CBERS-1) was successfully launched in 1999 as China's first-generation transmission-type Earth resource satellite. CBERS-02 was the successor to CBERS-01 and had the same function, composition, platform, payload, and nominal performance parameters as its predecessor. CBERS-02 was launched from the Taiyuan Satellite Launch Center on October 21, 2003.

The payload and orbital parameters of CBERS-01/2 are listed in Table 3.9 (China Center for Resource Satellite Data and Applications 2012; China Academy of Space Technology 2004). The CBERS-1/02 payload included three kinds of sensors: a charge-coupled device (CCD), an infrared multispectral scanner (IRMSS), and a

Table 3.9 Basic parameters of the CBERS-01/2 sensors

	CCD camera	Wide field imager (WFI)	Infrared multispectral scanner (IRMSS)
Type of sensor	Pushbroom	Pushbroom (discrete camera)	Oscillating scanning (forward and reverse)
Visible/near infrared band (μm)	1: 0.45–0.52 2: 0.52–0.59 3: 0.63–0.69 4: 0.77–0.89 5: 0.51–0.73	10: 0.63–0.69 11: 0.77–0.89	6: 0.50–0.90
Shortwave infrared band (μm)	N/A	N/A	7: 1.55–1.75 8: 2.08–2.35
Thermal infrared band (μm)	N/A	N/A	9: 10.4–12.5
Radiation quantization (bit)	8	8	8
Swath (km)	113	890	119.5
Number of pixels per band	5,812 pixels	3,456 pixels	Bands 6, 7 and 8: 1,536 pixels; band 9: 768 pixels
Spatial resolution (nadir) (m)	19.5	258	Bands 6, 7 and 8: 78 m; band 9: 156 m

Table 3.10 CBERS-02B technical parameters

Payload	Band no.	Spectral range (μm)	Resolution (m)	Swath (km)	Side view ability	Repetition period (d)	Data transmission rate
Panchromatic multispectral camera	B01	0.45–0.52	20	113	±32°	26	2 × 53
	B02	0.52–0.59	20				
	B03	0.63–0.69	20				
	B04	0.77–0.89	20				
	B05	0.51–0.73	20				
High-resolution camera (HR)	B06	0.5–0.8	2.36	27		104	60
Wide field imager (WFI)	B07	0.63–0.69	258	890		5	1.1
	B08	0.77–0.89	258				

wide field imager. Other loads included a high-density digital recorder (HDDR), a data collection system (DCS), a space environment monitor (SEM), and a data transmission system (DTS).

(2) *CBERS-02B*

CBERS-02B was an Earth observation satellite jointly developed by China and Brazil. The satellite was sent into orbit on September 19, 2007 from the Taiyuan Satellite Launch Center, and the first batch of Earth observation images was received on September 22, 2007. The satellite is no longer in operation. Its technical parameters are shown in Table 3.10.

CBERS-02B was equipped with three spatial resolutions: high, medium, and low. A combination of the CCD and HR images sent back from the satellite helped accurately identify and interpret residential areas, roads, forests, mountains, rivers, and other ground features. It could monitor the expansion of urban areas and provide a basis for urban planning and construction. Furthermore, it could provide support for decision making for precision agriculture. CBERS-02B could also be used to produce detailed maps such as dynamic land use maps and to update large-scale topographic maps.

(3) *ZY-1 02C*

The ZY-1 02C resource satellite was launched on December 22, 2011. It weighs approximately 2,100 kg and had a design life of three years. ZY-1 02C carries a panchromatic multispectral camera and a high-resolution panchromatic camera.

The satellite has two notable features. First, its 10-m resolution P/MS multispectral camera boasts the highest resolution of the multispectral cameras installed on China's civilian remote sensing satellites. Second, the two 2.36-m resolution HR cameras it carries make the monitoring swath as wide as 54 km, which greatly increased the data coverage and significantly shortened the satellite's repetition period. ZY-1 02C's payload parameters are shown in Table 3.11 (China Center for Resource Satellite Data and Applications 2012; China Academy of Space Technology 2004).

Table 3.11 ZY-1 02C sensor parameters

Parameter		P/MS camera	HR camera
Spectral range (μm)	Panchromatic	B1: 0.51–0.85	0.50–0.80
	Multispectral	B2: 0.52–0.59	
		B3: 0.63–0.69	
		B4: 0.77–0.89	
Spatial resolution (m)	Panchromatic	5	2.36
	Multispectral	10	
Width (km)		60	Single camera: 27; double camera: 54
Side view ability (°)		±32	±25
Repetition period (d)		3–5	3–5
Coverage period (d)		55	55

(4) *ZY-3*

The ZY-3 resource satellite was launched on January 6, 2012. It weighs approximately 2,650 kg and had a design life of five years. The satellite's mission is to continuously, reliably, and rapidly capture high-resolution stereo images and multispectral images of all parts of the country for a long period of time.

ZY-3 is China's first high-resolution civilian optical transmission-type stereo mapping satellite that integrates surveying, mapping, and resource investigation functions. The onboard front-view, rear-view, and vertical-view cameras can capture stereoscopic pairs in the same region from three different viewing angles to provide a wealth of three-dimensional geometric information. The image control and positioning precision are greater than one pixel. The swath of the front-view and rear-view stereoscopic pairs is 52 km wide and the baseline-height ratio is 0.85–0.95. The vertical image is 2.1 m, meeting the demand for 1:25,000 topographic map updates. ZY-3's payload parameters are shown in Table 3.12 (China Center for Resource Satellite Data and Applications 2012; China Academy of Space Technology 2004).

In 2012, ZY-3 sent back 1,590 batches of raw data, totaling 250 TB. The valid data covered 7.5 million square kilometers in China and 30 million square kilometers across the world. Imagery of Dalian, China, captured by the ZY-3 satellite is shown in Fig. 3.4.

3.2.3.2 Environment and Disaster Reduction Satellites

The environment and disaster reduction satellites are collectively referred to as the "China Small Satellite Constellation for Environment and Disaster Monitoring and Forecasting" ("Small Satellite Constellation" for short). The constellation is capable of using visible, infrared, microwave remote sensing and other means of observation

Table 3.12 ZY-3 sensor parameters

Platform	Payload	Band no.	Spectral range (μm)	Spatial resolution (m)	Width (km)	Side view ability (°)	Revisit time (d)
ZY-3	Front-view camera	–	0.50–0.80	3.5	52	±32	3–5
	Rear-view camera	–	0.50–0.80	3.5	52	±32	3–5
	Vertical-view camera	–	0.50–0.80	2.1	51	±32	3–5
	Multispectral camera	1	0.45–0.52	6	51	±32	5
		2	0.52–0.59				
		3	0.63–0.69				
		4	0.77–0.89				

Fig. 3.4 Image of Dalian, China, acquired by the ZY-3 satellite

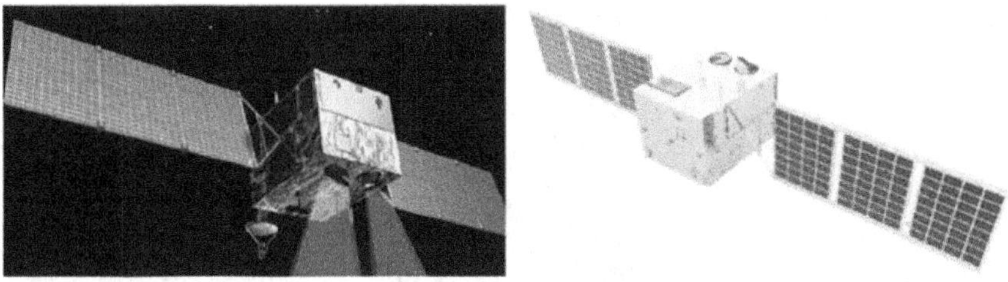

Fig. 3.5 The HJ-1A (left) and HJ-1B (right) satellites

to meet the needs of all-weather, 24-h observation and forecasting of natural disasters and environmental events.

(1) *HJ-1A/B*

The HJ-1A and HJ-1B environment and disaster reduction satellites were launched at 11:25 on September 6, 2008. HJ-1A carries a CCD camera and hyperspectral imager (HSI) and HJ-1B is equipped with a CCD camera and infrared scanner (IRS). HJ-1A and HJ-1B are equipped with the same type of CCD camera. The two cameras were placed symmetrically across the nadir, equally dividing the field of view. The cameras make parallel observations to achieve pushbroom imaging in four spectral bands with a 700-km Earth observation swath and a ground pixel resolution of 30 m. Additionally, the HSI on HJ-1A realizes pushbroom imaging in 110–128 spectral bands with a 50-km Earth observation swath and a ground pixel resolution of 100 m. HSI has a side view ability of ±30° and an onboard calibration function. The IRS on board HJ-1B completes imaging in four spectral bands (near, short, medium and long) with a 720-km Earth observation swath and a ground pixel resolution of 150/300 m. The two satellites are shown in Fig. 3.5.

(2) *HJ-1C*

HJ-1C is China's first S-band small SAR and environment and disaster reduction satellite, launched on November 9, 2012. HJ-1C has a mass of 890 kg and a sun-synchronous orbit at an altitude of 500 km. The local time of the orbital descending node is 18:00. Together with HJ-1A and HJ-1B, HJ-1C constitutes the first stage of China's environment and disaster reduction satellite constellation.

HJ-1C is equipped with an S-band SAR. Its payload works in two modes (strip mode and scanning mode) and employs a 6 × 2.8 m foldable mesh parabolic antenna. The SAR antenna was unfolded once HJ-1C entered orbit. It went into a swath imaging work mode after preparation. The onboard SAR has two imaging swaths: 40 and 100 km. The SAR's single-view spatial resolution is 5 m and the four-view spatial resolution is 20 m. Most of the HJ-1C's SAR images are taken in a multiview mode. The HJ-1C satellite is shown in Fig. 3.6.

The payload parameters of HJ-1C are shown in Table 3.13 (Satellite Environment Center, Ministry of Environmental Protection 2010a).

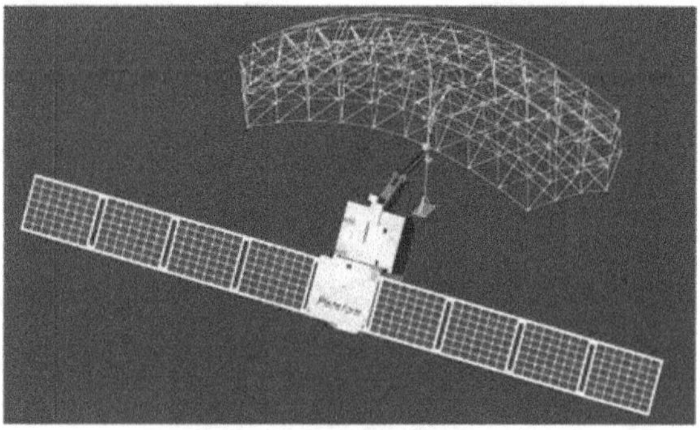

Fig. 3.6 The HJ-1C satellite

Table 3.13 HJ-1C's payload parameters

Parameter	Value
Operating frequency (MHz)	3,200
Side view	Side-looking
Spatial resolution (m)	5 m (single-view); 20 m (four-view)
Width of imaging swath (km)	40 (strip mode); 100 (scanning mode)
Radiometric resolution (dB)	3
Polarization mode	VV
Viewing angle (°)	25–47

3.2.3.3 Satellites of the High-Resolution Earth Observation Program

Globally, the United States was the first country to develop high-resolution Earth observation systems. Other countries such as Israel, France, and India have only one or two of these satellites each. Currently, China has no high-resolution satellites. According to the China Geographic Surveying and Mapping Information and Innovation Report (2012), although China has achieved success in satellite remote sensing technology, it is still behind in high-resolution civilian remote sensing satellite technology and its commercial applications.

GF-1 (Gaofen-1) was the first satellite of China High-resolution Earth Observation System (CHEOS) and was launched using an LM-2D rocket from the Jiuquan Satellite Launch Center on April 26, 2013. GF-1's development helped China master key technologies such as high spatial resolution, multispectral sensors, optical sensors, wide coverage, multipayload image mosaic fusion, precise and stable altitude control, and high-resolution data processing. Additionally, the development of GF-1 helped improve the capability for independent development of high-resolution satellites, and enhanced the self-sufficiency of high-resolution remote sensing data. The design life of GF-1 is five to eight years (Ding 2013).

Fig. 3.7 GF-2 image (resolution: 0.8 m)

On April 28, 2013, GF-1 began imaging and sending data. Data were received by the RADI Miyun Ground Station and processed by the China Center for Resource Satellite Data and Application. The first batch of images included four types: 2 m panchromatic, 8 m multispectral, 16 m multispectral, and 2 m panchromatic fused with 8 m multispectral.

GF-2 was launched successfully from Taiyuan Satellite Launch Center using an LM-4B carrier rocket on August 19, 2014. The successful launch was a result of special high-definition projects, indicating that Chinese remote sensing satellites were entering a submeter "high-definition era". GF-2's spatial resolution was 1.0 m and the swath width was 45 km, which was the largest imaging width of similar satellites of other countries (Fig. 3.7). GF-2 will be used for geographic and resource surveillance, environmental and climate change monitoring, precision agriculture, disaster relief, and city planning. The satellite is equipped with two cameras with the same resolution. The GF-2 camera can "twist its neck" to observe a range of ±35° in 180 s. GF-2 can swivel on its axis 35° to either side. Additionally, GF-2's five-year lifetime is longer than that of most other Chinese satellites, but the desired goal is eight years.

The GF-3 satellite is a new high-resolution SAR imaging satellite launched by an LM-4C rocket at 06:55 on August 10, 2016. It blasted off at the Taiyuan Satellite Launch Center in Taiyuan, the capital of northern China's Shanxi Province. As China's first C-band SAR imaging satellite that is accurate to one meter, it covers the globe with an all-weather, 24-h observation service and will be used for disaster warning, weather forecasting, water resource assessments, and the protection of maritime rights. With 12 imaging modes, the high-definition observation satellite can take wide pictures of the Earth and photograph detailed scenarios of specific areas. GF-3 is also China's first low orbit remote sensing satellite that has a lifespan

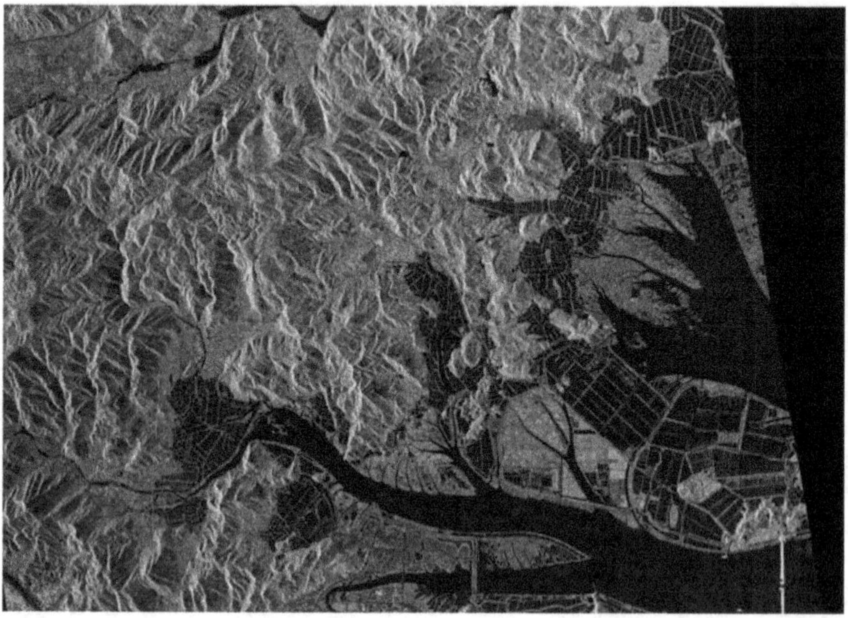

Fig. 3.8 GF-3 image (full polarization)

of eight years. It provides high-definition remote sensing data for its users over long periods of time. GF-3 is a polar orbit satellite with a high spatial resolution (Fig. 3.8) that can play a role in observing slowly changing objects such as water bodies, ice, and snow.

On June 26, 2015, China successfully launched the high-definition Earth observation satellite GF-8 into orbit from the Taiyuan Satellite Launch Center. GF-8 is an optical remote sensing satellite used in land surveying, urban planning, land delineation, highway and railway network design, crop yield estimation, disaster prevention and reduction, and other fields. The GF-9 satellite was launched from the Jiuquan Satellite Launch Center using an LM-2D carrier rocket on September 14, 2015. GF-9 is also an optical remote sensing satellite under CHEOS. The satellite can provide pictures with a ground pixel resolution of less than 1 m. It will be used in land surveying, urban planning, road network design, agriculture, and disaster prevention and relief.

On December 29, 2015, GF-4 was launched from the Xichang Satellite Launch Center in the southwestern province of Sichuan on board an LM-3B carrier rocket. It was the 222nd flight of the Long March rocket series. In contrast to GF-1 and GF-2, which orbit at low elevations (600–700 km) around Earth, GF-4 orbits 36,000 km away and moves synchronously with Earth. It can spot an oil tanker at sea using the CMOS camera, and features the best imaging capability among global high-orbit remote sensing satellites. GF-4 is China's first geosynchronous orbit HD optical imaging satellite and the world's most sophisticated HD geosynchronous orbit remote

Table 3.14 GF satellite parameters

Satellite	Sensor
GF-1	2 m panchromatic/8 m multispectral/16 m wide-swath multispectral
GF-2	1 m panchromatic/4 m multispectral
GF-3	1 m C-SAR
GF-4	50 m stationary gazing camera
GF-5	Visible shortwave infrared hyperspectral camera Full-spectrum spectral imager Atmospheric aerosol multiangle polarization detector Atmospheric trace gas differential absorption spectrometer Main atmospheric greenhouse gas monitor Ultrahigh-resolution infrared atmospheric sounder
GF-6	2 m panchromatic/8 m multispectral/16 m wide-swath multispectral
GF-7	High space three-dimensional mapping instrument

sensing satellite. It will be used for disaster prevention and relief, surveillance of geological disasters and forest disasters, and meteorological forecasting.

The GF-5 and GF-6 satellites were launched on May 9 and June 2, 2018, respectively. GF-5 was designed to run on a sun-synchronous orbit and carries six payloads: an advanced hyperspectral imager (AHSI), a visual and infrared multispectral imager (VIMI), an atmospheric infrared ultraspectral sounder (AIUS), a greenhouse gases monitoring instrument (GMI), an environmental trace gases monitoring instrument (EMI), and a directional polarization camera (DPC). The GF-6 satellite has a similar function to the GF-1 satellite but has better cameras, and its high-resolution images can cover a large area of the Earth, according to the State Administration of Science, Technology and Industry for National Defence. GF-6 can observe the nutritional content of crops and help estimate the yields of crops such as corn, rice, soybeans, cotton and peanuts. Its data will also be applied in monitoring agricultural disasters such as droughts and floods, evaluation of agricultural projects and surveying of forests and wetlands.

Parameters of the GF satellites are shown in Table 3.14.

3.2.3.4 Remote Sensing Microsatellites

Microsatellites are a new type of satellite that is low-cost and has a short development time and more flexible operation than conventional spacecraft that are heavy, costly, and time-consuming to develop. The spatial and temporal resolutions of Earth observation can be significantly improved using a distributed constellation of microsatellites. As a result, microsatellites are becoming more widely used around the world. China has launched several series of microsatellites for Earth observation, such as the "SJ" series, "Tsinghua-1", "NS-2", and "Beijing-1", which have improved and enriched the Chinese satellite observation system.

SJ-9A and SJ-9B are a new generation of microsatellite launched in 2012. They are the first satellites in the "New-tech Civilian Experimental Satellite" series. SJ-9A is equipped with a high-resolution multispectral camera with a panchromatic resolution of 2.5 m and multispectral resolution of 10 m. SJ-9B carries longwave infrared focal plane components for optical imaging with a resolution of 73 m. As of August 2013, the "SJ" satellite series had developed up to SJ-11E and provided adequate services for China's space science and technology experiments (Guo et al. 2013).

3.2.3.5 Remote Sensing from the Shenzhou Spacecraft

China has successfully developed and launched ten Shenzhou spacecraft, representing the country's achievements and capability in space science and technology. A series of scientific experiments such as space measurement, environmental monitoring, and Earth observation have been carried out in space with the support of the Shenzhou spacecraft. The Shenzhou spacecraft have accelerated the development of Earth observation technology in China.

In 2011, China's first space laboratory, Tiangong-1, was successfully launched. It was the starting point for Chinese space station development and signified that China had the ability to build short-term untended space stations. In the same year, Tiangong-1 successfully docked with the Shenzhou-8 unmanned spacecraft, revealing that China had achieved a series of key technologies such as space rendezvous and docking and operation of combined bodies. Shenzhou-9 and Shenzhou-10 were launched in 2012 and 2013, respectively. Shenzhou-11 was launched on October 17, 2016. For the first time, China realized space rendezvous and docking of manned spacecraft, and Chinese astronauts carried out teaching activities in space, marking an important step forward in China's space laboratory development (Jiang 2013). Figure 3.9 shows the development timeline of the Shenzhou series of spacecraft.

3.2.3.6 Commercial Remote Sensing Satellites

China's government is encouraging more participation from the private sector in commercial space programs to ensure the sustainable growth of the nation's space industry, and some commercial remote sensing satellites and missions have been launched or are planned, including Jilin-1, Beijing-2, SuperView-1, and Lishui-1.

The Jilin-1 satellites are China's first self-developed remote sensing satellites for commercial use and were launched from the Jiuquan Satellite Launch Center in northwestern China's Gansu Province on Oct. 7, 2015. The system includes one optical remote sensing satellite, two satellites for video imaging and another for testing imaging techniques. Jilin Province is one of China's oldest industrial bases and is developing its satellite industry as a new economic driver. The Jilin-1 GP 01 and 02 satellites for multispectral imaging were launched on a Long March 11 rocket from the Jiuquan Satellite Launch Center on January 21, 2019. By 2020, the plans

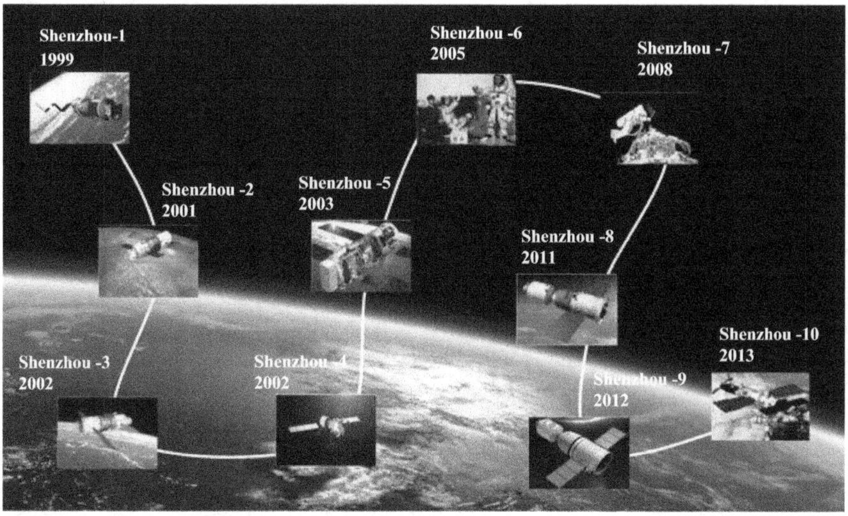

Fig. 3.9 Roadmap of the Shenzhou spacecraft program

indicate a 60-satellite orbital constellation capable of a 30-min update. From 2030, the Jilin constellation will have 138 satellites in orbit, forming a 24-h, all-weather, full-spectrum acquisition segment with the capability of observing any arbitrary point on the globe with a 10-min revisit capability, providing the world's highest spatial and temporal resolution space information products.

The Beijing-2 remote sensing satellite constellation comprises three identical optical EO satellites, which makes it possible to target any place on Earth once per day. The constellation provides less than 1-m high-resolution imagery products with a 23.4-km swath. The constellation was launched on July 10, 2015 from the Dhawan Space Centre in Sriharikota, India. The space and ground segments were designed to efficiently deliver timely information. The satellites were developed by the UK-headquartered Surrey Satellite Technology Ltd. (SSTL), which is the world's leading small satellite company and part of the Airbus Group. The Twenty First Century Aerospace Technology Company (21AT) will manage the satellites' operation, which includes observation and control, data reception and production, and related services. The satellites will provide the best combination of spatial resolution and temporal resolution to stimulate monitoring applications such as urban planning and intelligent management at a very high resolution. The main parameters of the constellation are shown in Table 3.15.

The SuperView-1 01 and 02 satellites were launched by one rocket on December 28, 2016, and two better performing satellites will be launched in the future to comprise four 0.5-m resolution satellites phased 90° from each other on the same orbit to provide services to global clients.

The Lishui-1 satellites, developed by the privately owned Zhejiang Lishui Electronic Technology Co Ltd, are commercial remote sensing satellites that were

Table 3.15 Parameters of the Beijing-2 satellite constellation

Feature	Parameter
Number of satellites	3
Satellite orbit	Sun-synchronous orbit Altitude: 651 km LTAN: 10:30
GSD	<1 m PAN <4 m MS
Bands	B/G/R/NIR
Swath width	23.4 km
MTF	PAN: 10% MS: 20%
Signal-to-noise	>100
Off-pointing capacity	±45°
Revisit	1 day
Lifetime	7 years

launched by an LM-11 solid-fuel rocket from the Jiuquan Satellite Launch Center in northwest China on November 10, 2016. The company plans to build a constellation of up to 80 to 120 commercial satellites to obtain images of the Earth and data to serve business purposes.

3.2.4 Other Land Observation Satellites

3.2.4.1 Japan's Satellites

In 1992, Japan's first Earth resource satellite, JERS-1, was launched into orbit. It carried next-generation SAR and optical sensors with a ground resolution of 18 m. During satellite operation, SAR transmits more than 1,500 microwave pulse signals per second to the surface and receives signals reflected from the ground with the same antenna. The optical sensor is composed of a VNIR radiometer and a shortwave infrared radiometer, and Earth observation is carried out in eight wavebands. Japan's Advanced Earth Observation Satellite (ADEOS), launched on August 17, 1996, was a next-generation large-scale Earth observation satellite that followed Japan's marine observation satellite, MOS, and Japan's Earth resource satellite, JERS-1.

On January 24, 2006, the Japan Space Agency launched the ALOS-1 satellite. ALOS-1 used advanced land observation technologies to obtain flexible, higher resolution Earth observation data that could be applied to mapping, regional observation, disaster monitoring, resource surveys, technical development, and other fields. The basic parameters of the ALOS-1 satellite are shown in Table 3.16.

The JAXA completed operation of ALOS-1 on May 12, 2011. The technologies acquired from ALOS-1 operation were succeeded by the second Advanced Land Observing Satellite (ALOS-2). The PALSAR-2 on board ALOS-2 is an L-band SAR

Table 3.16 ALOS-1 characteristics

Parameter	Value
Launch date	2006.01.24
Type of orbit	Sun-synchronous orbit
Repetition period (d)	46
Altitude (km)	691.65
Inclination (°)	98.16
Attitude control precision (°)	2.0×10^{-4} (in coordination with ground control point)
Positioning accuracy (m)	1.0
Data rate (Mbps)	240 (via data relay satellites)
Onboard data storage	Solid-state data recorder (90 GB)

sensor, a microwave sensor that emits L-band radio waves and receives their reflection from the ground to acquire information. The PALSAR-2 has three modes: (1) Spotlight mode—the most detailed observation mode with 1 by 3 m resolution (25 km observation width); (2) Strip Map mode—a high-resolution mode with the choice of 3, 6 or 10 m resolution (observation widths of 50 or 70 km); and (3) ScanSAR mode—a broad area observation mode with observation widths of 350 or 490 km and resolution of 100 or 60 m, respectively.

3.2.4.2 India's Satellites

Resourcesat is part of the Indian remote sensing satellite system. The first of the Resourcesat satellites, Resourcesat-01, was launched on October 17, 2003. This series is used for disaster forecasting, agriculture, water resources, forest and environment monitoring, infrastructure development, geological exploration, and mapping services.

The second satellite of this series, Resourcesat-02, was the 18th remote sensing satellite designed and developed by ISRO (Fig. 3.10). With a total mass of 1,206 kg, Resourcesat-02 adopts three-axis stabilization technology and was designed to work for five years. Its sensors and related subsystems were jointly developed by the ISRO Satellite Center (ISAC) and the Space Application Center (SAC). The Indian National Remote Sensing Center (NRSC) is responsible for receiving and preprocessing the satellite's image data as well as for production and distribution of products. Resourcesat-02 enhanced the Earth observation capability of the country's remote sensing satellite system to better serve India's economic development and national defense.

Resourcesat-02 replaced Resourcesat-01 after a series of on-orbit tests, and expanded ISRO's remote sensing data services. The Resourcesat-02 satellite's payload includes: linear imaging self-scanning sensors (LISS-3 and LISS-4), an

Fig. 3.10 The Resourcesat-02 satellite

advanced wide field-of-view sensor (AWIFS), three high-resolution multispectral cameras, and a marine automatic identification system (AIS). LISS-4 has a spatial resolution of 5.8 m and scanning width of 70 km, can work in the VNIR spectral range, and can obtain cross-track stereo images (Goward et al. 2012).

3.2.4.3 Russia's Satellites

The Resurs-F series of satellites are tasked with monitoring crop growth, ice cover, landforms, and other features. They also undertake scientific research missions. For example, the two Resurs-F1 satellites launched in May and July 1989 were passive atmospheric research satellites, 70 mm in diameter and 78 kg in mass, that were used to study the density of the upper atmosphere. The two satellites also carried scientific instruments from other countries for scientific experiments.

The first Resurs-F satellite was launched on September 5, 1979 from the Plesetsk Launch Site using an SL-4 rocket. The satellite was 7 m long, 2.4 m in diameter, 6,300 kg in mass and was composed of three compartments. The central part of the satellite was a 2.3-m diameter sphere that housed the imaging system, electronic control system, and recovery system. One side was connected to the 3 m long and 2 m wide propulsion module via a fixing mechanism that unlocked when the retarding rocket was ignited. The other side was 1.9 m long and the propulsion unit occupied up to 1.0 m. The propulsion unit was used for orbital adjustment and was cast off when the return capsule re-entered the atmosphere. The remaining 0.9 m of space was

Fig. 3.11 The Resurs-F1 satellite

used to carry additional releasable secondary payloads of up to 30 kg or more. These secondary payloads could be placed inside or outside the return capsule and carried back to the ground. An overview of the Resurs-F1 satellite is shown in Fig. 3.11.

The imaging system on board the Resurs-F1 satellite included an SA-20M long-focus wide imaging system with a KFA-1000 camera and an SA-34 wide mapping and imaging system with a KATE-200 camera. Compared with Resurs-F1, the Resurs-F2 satellite's biggest improvement is the addition of two solar panels, which extended its service life to nearly one month. The first Resurs-F2 satellite, also known as Cosmos-1906, was launched into space in 1987. However, the launch was unsuccessful and the satellite was destroyed in orbit. Resurs-F2 satellites are operating in 170–240 km low Earth orbits and near-polar circular orbits with an orbital inclination of 82.3°. An outline of the Resurs-F2 satellite is shown in Fig. 3.12.

The Resurs-F2's imaging system is significantly different than that of Resurs-F1 and includes a KFA-1000 camera and a high-resolution MK-4 mapping camera. Equipped with a passive remote sensor, the MK-4 camera can record images on three separate pieces of film and perform imaging in any three of the following six spectral bands: 0.63–0.69 μm, 0.81–0.90 μm, 0.52–0.57 μm, 0.46–0.51 μm, 0.58–0.80 μm, and 0.40–0.70 μm. The camera's focal length is 300 m, the spatial resolution is greater than 10 m, the panchromatic spectral resolution is 8 m, and the ground width is 120–180 km. One scan can generate 2,700 images and the image size is 180 × 180 mm with an overlap ratio of 60%. The satellite can be used for mapping, environmental monitoring, and geographic surveys.

The Resurs-O series of satellites were mainly used in geology, cartography, fire detection, ice detection, hydrology, and agriculture. They were designed and manufactured by the then National Institute of Electronics in the former Soviet Union.

Fig. 3.12 The Resurs-F2 satellite

3.3 Ocean Observation Satellites

Ocean satellites are the best tools for understanding Earth's oceans, and can be economically used for real-time, synchronous, and continuous monitoring of large areas. At present, ocean satellites are the primary means of marine environment monitoring, making their development a necessity. Ocean satellites can enhance scientists' capability for marine environment and disaster monitoring, forecasting, and early warning, and can provide efficient services for marine resource surveying, development, and management. These satellites can conduct global surveys of fisheries, scientifically estimate fishery potential, and provide a basis for the development of fishery policies. Furthermore, they can effectively and affordably measure the marine gravity field to provide an understanding of submarine tectonics and oil and gas reserves, and assist in developing offshore oil fields.

3.3.1 *US Ocean Observation Satellites*

3.3.1.1 Development Stages of US Ocean Satellites

The development of US ocean satellites has experienced four stages (Dong 2012): (1) preparation stage (before 1978); (2) experiment stage (1978–1985); (3) application research stage (1985–1999), and (4) comprehensive oceanographic observation stage (1999–present).

(1) Preparation stage

The first US meteorological satellite, TIROS-I, was launched by NASA in April 1960, followed by TIROS-II, which started sea surface temperature observation. In 1961, the United States began to implement the Mercury Program, making it possible for astronauts to observe the ocean from a high altitude. In 1969, NASA began to promote a marine observation plan; in 1975, GOES-3 was equipped with an altimeter for measuring the distance from the satellite to the sea surface. In 1973, the Skylab space station confirmed the potential of visible and infrared remote sensing in continuous Earth observation.

(2) Experiment stage

In this stage, marine remote sensors were mainly installed on US ocean satellites such as Seasat, Nimbus-7, TIROS-N, and GEOS. The main marine elements inversed in this stage included sea surface temperature, ocean color, and sea ice. In 1981, NOAA satellites began using the multichannel sea surface temperature (MCSST) algorithm to forecast sea surface temperature.

(3) Application research stage

The main ocean satellites launched in this stage were equipped with a variety of microwave monitoring instruments, infrared radiometers and ocean color imagers to monitor the sea surface, submarine topography, sea waves, sea wind, ocean currents, marine pollution, primary oceanic productivity, and other factors. In 1985, the United States launched an ocean topography satellite called Geosat, which was mainly used to measure significant wave height, wind velocity and meso-scale oceanic features. Over the years, Geosat provided a wide range of altimeter data. Other meteorological satellites were also involved in marine observation. For instance, NOAA meteorological satellites were used for sea surface temperature inversion, sea condition monitoring, and sea pollution research. In 1987, the SeaWiFS Working Group of NASA and the Earth Observation Satellite Company (EOSAT) jointly proposed a systematic plan for spaceborne wide-field-of-view marine observation. In August 1997, the United States launched an ocean satellite, SeaSTAR, (also called OrbView-2), which was later included in the EOS program as the first ocean color satellite of the program. Subsequently, the United States developed the navy remote ocean sensing system (NPOSS) and, in cooperation with France, NASA developed TOPEX/Poseidon for observing ocean topography.

(4) Comprehensive oceanographic observation stage

According to the research objectives of the EOS and ESE, the period from 1999 to the present is the comprehensive oceanographic observation stage in the development of ocean remote sensing. The first satellite of the next-generation international Earth observation satellite system, Terra (EOS-AM1), was launched on December 18, 1999, marking the beginning of a new era of human observation of Earth. The second polar-orbiting environmental remote sensing satellite, Aqua (EOS-PM1), was launched on May 4, 2002. Both Terra and Aqua are equipped with a Moderate Resolution Imaging Spectroradiometer (MODIS) that has 36 wavebands ranging from visible to thermal infrared light, nine of which can be used for ocean color remote sensing. Compared with SeaWiFS, MODIS is more advanced and is known as the third-generation ocean color (and meteorological element) sensor (DeVisser 2013). The Jason program was proposed to meet the requirements for establishing a global marine observation system and the demands of oceanic and climatological research. The Jason-2 ocean altimetry satellite (also used for accurate determination of ocean topography) was jointly developed by the Centre National d'Etudes Spatiales (CNES), EUMETSAT, NASA, and NOAA and launched on June 20, 2008. As a follow-up to TOPEX/Poseidon and Jason-1, it is an important observation platform for global oceanographic studies.

3.3.1.2 Typical US Ocean Satellite Systems

(1) Seasat-1

Launched on June 27, 1978, Seasat-1 operated on orbit for 105 days and stopped working on October 10, 1978, due to an electrical system fault. It was launched to demonstrate global monitoring technologies including the observation of oceanic dynamics and satellite orbit characteristics and to provide oceanographic data for the development and application of an operational ocean dynamics monitoring system.

Seasat-1 was the first ocean satellite to use synthetic aperture radar (SAR) for ocean observation by means of remote sensing (Fig. 3.13). Its purpose was to prove the feasibility of using satellites to monitor global oceanic phenomena and help determine the requirements of ocean remote sensing satellite systems. The goal was to collect data about ocean surface wind, sea surface temperature, atmospheric water, sea ice characteristics, ocean topography, and similar parameters. Seasat-1 could cover 95% of the world in a 36-h observation cycle.

(2) OrbView-2

Also called SeaStar, OrbView-2 was launched into a 705 km sun-synchronous orbit on August 1, 1997. The mass of the parent capsule was 155 kg, the mass of the instruments was 45.4 kg, and that of the satellite was 317 kg. The outer dimensions of the satellite were 1.15 × 0.96 × 1.6 m, and the solar wing plate had a span of 3.5 m when unfolded (Fig. 3.14).

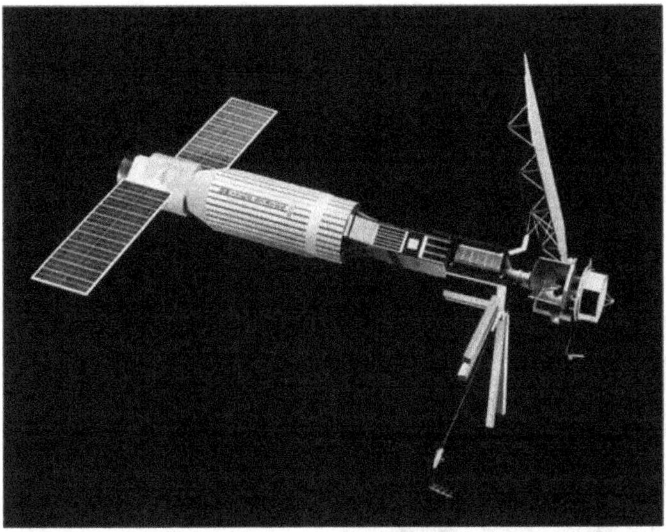

Fig. 3.13 Seasat-1

The satellite carried only one remote sensing instrument, SeaWiFS, which could monitor ocean color, generate multispectral images of the land and sea surface, and analyze the impacts of ocean color changes on the global environment, atmosphere, carbon cycle, and other ecological cycles. SeaWiFS consisted of optical remote sensors and an electronic module, and the satellite covered the global ocean area once every two days.

OrbView-2 was the world's first satellite that could generate color images of the Earth every day. The imager had eight spectral segments, six of which were visible and two of which were near infrared. With a spatial resolution of 1.1 km and a 2,800 km scanning width, OrbView-2 data could be used in the fishing industry, agriculture, scientific research, and environmental monitoring.

(3) Jason-1

As an ocean satellite, Jason-1 is used to study the relationship between the ocean and the atmosphere, monitor global ocean circulation, improve global weather prediction and forecasting, and monitor El Niño, ocean eddies, and other events (Chander et al. 2012). With a total weight of 500 kg and payload of 120 kg, Jason-1 was launched on December 7, 2001 (Fig. 3.15). It was the world's first satellite to use the French Alcatel PROTEUS multifunctional microplatform and carried five scientific instruments: one dual-frequency solid-state spaceborne radar altimeter (Poseidon-2), which was the main payload of Jason-1, one triple-channel microwave radiometer (JMR) used to measure atmospheric water vapor content and provide water vapor correction for the radar altimeter, and three other instruments for accurate orbit determination that comprise one Doppler orbitography by radio positioning integrated by satellite

Fig. 3.14 OrbView-2

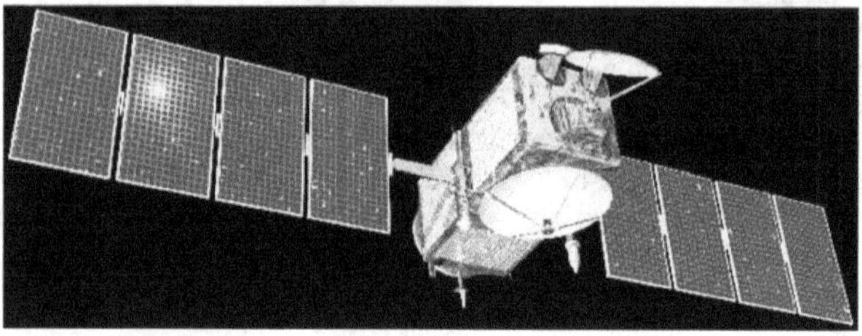

Fig. 3.15 Jason-1 ocean satellite

(DORIS), one laser retro reflector array (LRA), and one turbo rogue space receiver (TRSR).

As the main payload of the Jason-1 satellite, Poseidon-2 was developed by the CNES as an improved model of the Poseidon-1 radar altimeter. In addition to inheriting all the advantages of its predecessor, Poseidon-2 used dual-frequency technology, with working frequencies of 13.575 GHz (Ku-band) and 5.3 GHz (C-band). Compared to other radar altimeters, Poseidon-2 was smaller in volume and lighter weight and had more efficient power consumption. It is mainly used to measure sea surface height, wind velocity, significant wave height, and ionospheric corrections. The main technical parameters of the Poseidon-2 radar altimeter are shown in Table 3.17.

Table 3.17 Main technical parameters of the Poseidon-2 radar altimeter

Satellite feature	Parameter
Operating frequency (GHz)	13.575 (Ku), 5.3 (C)
Pulse repetition frequency (PRF) (Hz)	2,060
Pulse duration (μs)	105
Bandwidth (MHz)	320
Antenna diameter (m)	1.2
Antenna wave width (°)	1.28 (Ku), 3.4 (C)
Power (W)	7

3.3.2 European Ocean Observation Satellites

The successful launch of the first meteorological satellite, Meteosat, in 1977 marked the beginning of the implementation of the European Earth Observation Program (EOP). The main task of Meteosat was to monitor the atmosphere over Europe and Africa. Implementation of the ERS missions in the early 1990s marked the EOP's entry into a new stage. The launch of an ENVISAT satellite in 2002 sped up the pace of EOP implementation. The ESA proposed the Living Planet Programme (LPP) in 1998. Compared with the ERS and ENVISAT missions, the LPP used smaller satellites, was less costly and had better defined targets.

3.3.2.1 ERS-1/2

The ERS-1/2 satellites operated on a near-polar sun-synchronous orbit, with an average orbital altitude of 785 km and an orbital inclination of 98.50°. The local time when the satellite moved from north to south across the equator was 10:30 AM. The ERS-1 launch involved a number of adjustments to the orbital altitude instruments. The three months after launch, the satellite used a three-day period for trial operation at an orbital altitude of 785 km (reference orbit). The orbital adjustment period of the sun-synchronous satellite was 3–176 days, and the main working period was 35 days. The average orbital altitude for the three-day period was 785 km, the orbital altitude above the equator was 909 km, and the satellite circled Earth 43 times. The main parameters of ERS-1/2 are shown in Table 3.18.

The satellite platform carried the following seven instruments (Fig. 3.16): (1) an active microwave instrument (AMI) with an SAR that had a 100-km mapping swath; (2) a wind scatterometer that used three groups of antennas to measure the direction and velocity of sea surface winds; (3) a radar altimeter that was used to accurately measure sea surface topography and elevation, wave height, sea surface wind velocity, and characteristics of sea ice; (4) an orbit-tracking scanning radiometer and microwave sounder; (5) a precision ranging velocimeter that was used to accurately measure the satellite position, orbital characteristics, and the position of fixed ground stations; (6) a laser reflector that used laser beams emitted from the

Table 3.18 ERS-1/2 parameters

Satellite parameter	Value
Weight (kg)	2,400
Total length (m)	11.8
Solar cell array	Area: 11.7 × 2.4 m^2; power: 1.8 KW; service life: 2 years
SAR antenna (m)	10 × 1
Scatterometer antenna (m^2)	Anterio-posterior direction: 3.6 × 0.25; middle direction: 2.3 × 0.35
Radar altimeter antenna diameter (m)	1.2
Communication frequency band	S-band
Orbit	800 km sun-synchronous orbit
Orbital period (d)	35

Fig. 3.16 ERS-1

ground station to measure the satellite orbit and position; and (7) an onboard data processing system.

3.3.2.2 ENVISAT Satellite

Launched on March 1, 2002, ENVISAT was a polar-orbiting Earth observation satellite and the largest Earth observation satellite built (Fig. 3.17). ENVISAT had ten

Fig. 3.17 ENVISAT satellite

Table 3.19 The working modes and characteristics of the ASAR sensor on the ENVISAT satellite

Feature	Image	Alternating Polarization	Wide Swath	Global Monitoring	Wave
Imaging swath width (km)	Max. 100	Max. 100	Approx. 400	Appr. 400	5
Downlink data rate (Mbps)	100	100	100	0.9	0.9
Polarization mode	VV or HH	VV/HH or VV/VH or HH/HV	VV or HH	VV or HH	VV or HH
Resolution (m)	30	30	150	1,000	10

instruments that constituted an observation system that captured lithosphere, hydrosphere, atmosphere, biosphere, and ice layer information.

At the time, the ASAR on board ENVISAT was the world's most advanced spaceborne SAR sensor with new features including multipolarization, multiple modes, and multiple incident angles. The ground resolution of data reached 25 m, and the widest coverage was 400 km. The multipolarization SAR imaging system could acquire copolarization and cross-polarization information of ground objects and more accurately detect features of a target. The five working modes and characteristics of the ENVISAT satellite's ASAR sensor are listed in Table 3.19.

3.3.2.3 The Gravity Field and Steady-State Ocean Circulation Explorer (GOCE)

The GOCE was a satellite that adopted new technologies to map the Earth's gravitational field (Fig. 3.18). The GOCE was launched on March 17, 2009 (Metzler and Pail 2005). The satellite started scientific observation activities on September 30, 2009 and carried out its functions during its service life. In October 2010, the first batch of GOCE satellite data was released freely to scientific researchers and noncommercial users across the world, opening up a new historical period for Earth gravity field research.

The GOCE moved on a low, nearly-circular, twilight sun-synchronous orbit. The orbital plane's eccentricity was less than 0.001 and its inclination was 96.7°, leaving a nonobservable area with a spherical radius of approximately 6.7° in the northern and southern polar regions. The satellite's working time was twenty months, including three months of commissioning and calibration followed by a period of scientific measurement and period of dormancy. Due to its energy supply, trial operation, gradiometer calibration, orbital adjustment and other reasons, the time period for scientific observation was only twelve months. Once the satellite's working time period had expired, it was decided to extend the GOCE's operational period based on the working state of all systems and the quality of data products obtained. The original plan was to extend the mission by ten months and increase the observation tasks accordingly (Floberghagen et al. 2011).

The goal of the GOCE mission was to provide a high-precision, high-resolution static Earth gravity model (Bouman et al. 2009). Such models can be obtained based

Fig. 3.18 GOCE

on the gravity gradient and GPS tracking data. The specific goals were to: determine global gravity anomalies with a precision of 1 mGal, determine the global geoid with a precision of 1–2 cm, and fulfil these goals with a spatial resolution above 100 km (half-wavelength) (Visser 2010; Gooding et al. 2007).

3.3.3 China's Ocean Observation Satellites

China's first independently developed ocean satellite, HY-1A, was launched on May 15, 2002. As an experimental satellite, HY-1A was used to monitor ocean color and temperature. HY-1B was launched on April 11, 2007, and was positioned for operation on September 3. HY-1B was the successor to HY-1A, with a design life of three years, and its technical indicators and functions were superior to those of HY-1A. The HY-2A satellite was launched on August 16, 2011. As a marine dynamic environment satellite, HY-2 worked to detect the sea surface wind field, temperature field, sea surface height, wave field, and flow field. It adopted the platform of the ZY-1 satellite. A roadmap of ocean satellite development is shown in Fig. 3.19.

(1) HY-1A

The ten-band Chinese ocean color and temperature scanner (COCTS) was used to detect ocean color environmental factors (concentration of chlorophyll, content of suspended sediments, and presence of soluble organic matter) and temperature field. The satellite had a nadir ground resolution of 1,100 m, 1,024 pixels per line,

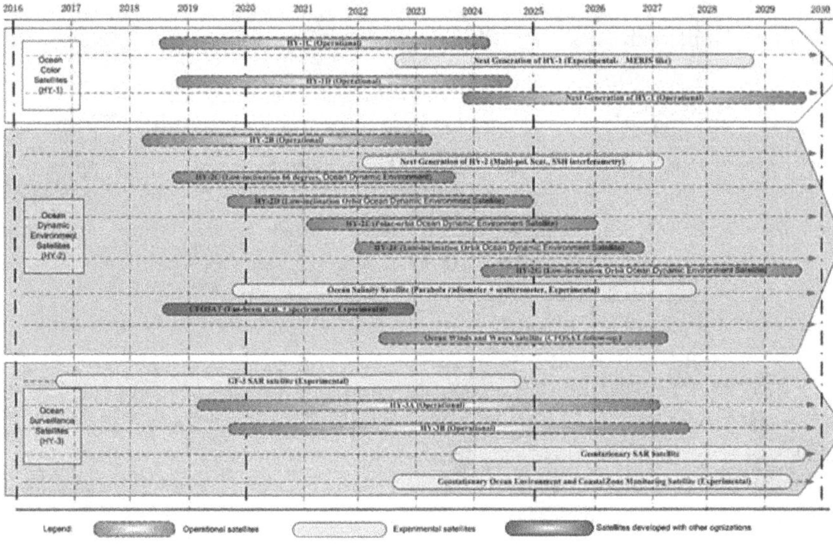

Fig. 3.19 Roadmap of ocean satellite development

Table 3.20 Ocean color and temperature scanner parameters

Parameter	Value
Spectral range (μm)	B1: 0.402–0.422, B2: 0.433–0.453 B3: 0.480–0.500, B4: 0.510–0.530 B5: 0.555–0.575, B6: 0.660–0.680 B7: 0.740–0.760, B8: 0.845–0.885 B9: 10.30–11.40, B10: 11.40–12.50
Band-center wavelength shift (nm)	≤2(B1-B8)
Nadir ground resolution (m)	≤1100
Number of pixels per line	1664
Quantization level (bit)	10
Radiometric precision	Visible light: Infrared: ±1 K (when the onboard calibration accuracy is 300 K)

a quantization level of 10 bits, and a radiometric precision of 10% of the visible light. The four-band CCD imager was used to monitor coastal zone dynamics to obtain relatively high-resolution images of land-sea interaction areas. The imager had a nadir ground resolution of 250 m, 2,048 pixels per line, and ≤5% degrees of polarization.

(2) HY-1B

As the successor of HY-1A, the HY-1B ocean satellite was launched on April 11, 2007, and had a design life of three years. Its payload parameters are shown in Table 3.20 (National Satellite Ocean Application Service 2007, 2011). HY-1B monitors the Bohai Sea, the Yellow Sea, the East China Sea, the South China Sea, and their coastal zones to detect chlorophyll, suspended sediments, soluble organic matter, and sea surface temperature.

(3) HY-2A

The HY-2A ocean satellite was China's first marine dynamic environment satellite to integrate active and passive microwave remote sensors and is capable of high-precision orbital measurement and determination, and all-weather, 24-h global detection. Its mission is to monitor and investigate marine environments and obtain dynamic ocean environment parameters including sea surface wind, wave height, ocean current, and sea surface temperature. HY-2A also provides data for the pre-warning and forecasting of disastrous sea conditions, and offers supportive services for the prevention and mitigation of marine disasters, protection of marine rights and interests, development of marine resources, protection of the marine environment, marine scientific research, and national defense. HY-2A was launched at 06:57 on August 16, 2011 from the Taiyuan Satellite Launch Center using a CZ-4B rocket.

The satellite is equipped with a scanning microwave radiometer, a radar altimeter, a microwave scatterometer, a calibrated microwave radiometer, DORIS, dual-frequency GPS, and a laser range finder. The parameters of the radar altimeter are shown in Table 3.21.

Table 3.21 Technical parameters of the HY-2 radar altimeter

Parameter	Value
Operating frequency (GHz)	13.58, 5.25
Pulse limited footprint (km)	≤2
Altitude measurement precision (cm)	<4
Effective wave height measurement range (m)	0.5–20

3.3.4 Other Ocean Observation Satellites

In addition to the United States and the ESA, Russia, Japan, Canada, and India have launched various ocean satellites into space. Generally, modern ocean satellites have an accurately determined sun-synchronous orbit, use a variety of remote sensors for measurement, and adopt a comprehensive remote sensing platform.

3.3.4.1 Japan's Satellites

On February 19, 1987, Japan launched its first ocean observation satellite, MOS-1, on an N-l rocket from the Tanegashima Space Center (Fig. 3.20).

MOS-1 was loaded with two optical remote sensors: a multispectral electronic self-scanning radiometer (MESSR) and a visible thermal infrared radiometer (VTIR). Other payloads included a microwave scanning radiometer (MSR), a data collection system (DCS), and a visible thermal infrared repeater. The MESSR is an electronic scanning optical observation remote sensor that uses a CCD to capture land and ocean information. Wavelengths ranging from visible light to near infrared are divided into four spectral bands (see Table 3.22). On board the satellite were two identical devices

Fig. 3.20 MOS-1

Table 3.22 MOS-1 sensor characteristics

	MESSR	VTIR		MSR
Observation purpose	Ocean color, land use, etc.	Sea surface temperature, etc.		Water vapor, ice, snow, etc.
Observed wavelength (μm)	0.51–0.59	0.5–0.7	6–7	
	0.61–0.69			
	0.72–0.80		10.5–11.5	
	0.80–1.1		11.5–12.5	
Instantaneous field of view (km)	0.05	0.9	2.7	32, 23
Radio wave resolution	39–15 dB	55 dB	0.5 K	<1.5 K
Observation width (km)	100	1500		317
Scanning mode	Electronic scanning	Mechanical scanning		Mechanical scanning

with a land observation width of 100 km, coordinated coverage of 185 km, and ground resolution of 50 m.

3.3.4.2 India's Satellites

OceanSat-1 was launched for the study of marine physics and marine biology on May 26, 1999 using a PSLV-C2 rocket (Dash et al. 2012). It was equipped with an ocean color monitor (OCM) and a multifrequency scanning microwave radiometer (MSMR) (Fig. 3.21). The OCM was used to collect data and worked at 402–422 nm, 433–453 nm, 480–500 nm, 500–520 nm, 545–565 nm, 660–689 nm, 745–785 nm, and 845–885 nm with a spatial resolution of 360 m and width of 1,420 km.

OceanSat-2 was launched on September 23, 2009 using a PSLV-C14 rocket. It functions on a circular near-polar sun-synchronous orbit 720 km above Earth, and continuously provides effective IRS-P4 services (Gohil et al. 2013; Sathiyamoorthy et al. 2012). The observation data from OceanSat-2 are applied to new areas of ocean research such as tornado trajectory prediction, coastal area mapping, and atmospheric research. The OCM and ROSA provide several geophysical parameters such as suspended sediment, yellow matter, phytoplankton, sea surface temperature (SST), sea wind, sea conditions, significant wave height, and atmospheric profiles derived from GPS radio occultation.

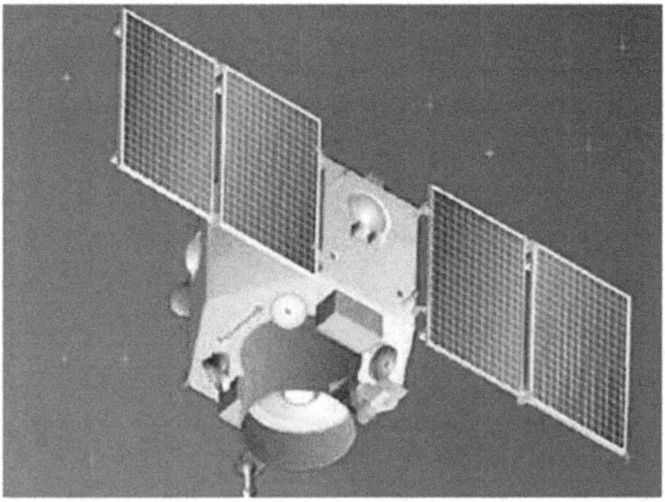

Fig. 3.21 OceanSat-1

3.3.4.3 Russia's Satellites

Since 1979, the Soviet Union/Russian Federation has launched a series of ocean color satellites known as the Okean-O1 series of satellites for marine and polar ice observation (Fig. 3.22). Twelve Okean-O1 satellites were launched (including one launch failure) by the end of August 1995 and four satellites were launched between May 1988 and October 1994, referred to as Okean-1 to Okean-4. The satellite payloads included an X-band side-looking radar with 350/1,500 m resolution and 1,380/1,930 km scanning width, and a microwave radiometer with an 8 mm working

Fig. 3.22 Okean-O1 Satellite

frequency and a 550 km scanning width. The Okean-O1 series of satellites functioned at an orbital altitude of 650 km and an inclination of 82.5°. Each satellite weighed 1.95 t and had a design life ranging from six months to a year. In 1999, Russia launched a new type of ocean satellite, Okean-O, whose design life and weight were increased to three years and 6.5 t, respectively. The Okean-O series of satellites adopted a sun-synchronous orbit with an altitude and inclination of 670 km and 98°, respectively. Each satellite was equipped with nine remote sensors, leading to improved optical resolution (25–200 m for visible light and 100–600 m for infrared).

3.3.4.4 Canada's Satellites

RADARSAT is a joint research project conducted by Canada (Canadian Space Agency/Canada Centre for Remote Sensing) and the United States (NASA). The radar is designed to provide detailed information for sea ice, land ice, and climate studies, and the radar images can be used in fields such as oceanography, agriculture, forestry, hydrology, geology, and geography and to provide real-time ice surveillance of the Arctic ocean.

RADARSAT-1 was launched by Canada on November 4, 1995 (Fig. 3.23). Satellite-borne SAR is an active remote sensing device. Because it actively emits electromagnetic waves to obtain information, it can penetrate clouds and fog and overcome night barriers and is capable of all-weather, 24-h observation. It can observe the surface on a regular basis and obtain real-time observation data. The SAR on board RADARSAT-1 was a C-band multiangle sensor with an HH polarization mode and

Fig. 3.23 RADARSAT-1

seven working modes used for coastal zone observation, sea ice monitoring, topographic surveys, and other uses.

RADARSAT-2 was launched in December 2007 as Canada's next-generation commercial radar satellite offering powerful technical advancements for mapping in Canada and around the world. This satellite is a follow-up to RADARSAT-1. It has the same orbit and is separated by half an orbit period (~50 min) from RADARSAT-1 (in terms of the ground track, this represents ~12 days of separation). RADARSAT-2 is a C-band imaging radar system, with a nominal imaging swath from 20 to 500 km, incidence angles from 10° to 60°, and fully polarimetric imaging capability; it is an indispensable tool for managing natural resources and monitoring the environment in the twenty-first century. It fills a wide variety of roles, including in sea ice mapping and ship routing, iceberg detection, agricultural crop monitoring, marine surveillance for ship and pollution detection, terrestrial defense surveillance and target identification, geological mapping, land use mapping, wetlands mapping, and topographic mapping.

3.4 Meteorological Observation Satellites

Meteorological satellites have become an indispensable part of the basic and strategic resources for national economic and social development in countries across the world. As the problems of environmental pollution, resource shortages, and natural disasters become increasingly worse, the role of meteorological satellites in weather forecasting, environmental monitoring, and disaster mitigation and prevention has become more important than ever.

3.4.1 US Meteorological Observation Satellites

Since the launch of its first meteorological satellite in April 1960, the United States has developed two series of meteorological satellites: geostationary meteorological satellites and polar-orbiting meteorological satellites. The former is the Geostationary Operational Environmental Satellite (GOES) series and the latter comprises NOAA satellites in the Defense Meteorological Satellite Program (DMSP).

3.4.1.1 The DMSP Satellite System

DMSP satellites operate on sun-synchronous orbits. Some of the orbital parameters are listed in Table 3.23.

The DMSP satellite series adopts a double-satellite operation system. One satellite operates on a 06:00 AM orbit and the other on a 10:30 AM orbit, both with a repeat observation cycle of twelve hours and seven payloads, which are shown in Table 3.24.

Table 3.23 Orbits of the current DMSP system satellites

Satellite code	Orbital altitude (km)	Orbital period (min)	Orbital inclination (°)	Launch time	Orbiting direction
DMSP 5D 3/F14	833	101	98.7	20:29	Clockwise
DMSP 5D 3/F15	833	101	98.9	20:29	Clockwise
DMSP 5D 3/F16	833	101	98.9	21:32	Clockwise
DMSP 5D 3/F17	850	101	98.7	17:31	Clockwise
DMSP 5D 3/F18	850	101	98.7	17:31	Clockwise

Table 3.24 Payloads of the DMSP system satellites in orbit

Satellite code	Payloads
DMSP 5D 3/F14	OLS, SSB/X-2, SSI/ES-2, SSJ/4, SSM, SSM/I, SSM/T-1, SSM/T-2
DMSP 5D 3/F15	OLS, SSI/ES-2, SSJ/4, SSM, SSM/I, SSM/T-1, SSM/T-2
DMSP 5D 3/F16	OLS, SSI/ES-3, SSJ/5, SSM, SSM/IS, SSULI, SSUSI
DMSP 5D 3/F17	OLS, SSI/ES-3, SSM, SSM/IS, SSULI, SSUSI
DMSP 5D 3/F18	OLS, SSI/ES-3, SSM, SSM/IS, SSULI, SSUSI

The DMSP satellite series uses two data transmission modes: direct reading mode and storage mode. The former can transmit data to the ground station in real time and the latter transmits the data stored in the satellite-borne magnetic tape unit to the ground station when the satellite is flying over it. These ground stations include the Fairchild Air Force Base in the state of Washington, the Loring Air Force Base in Maine, and the Ka'ena Point Satellite Tracking Station in Hawaii. Then, the ground stations transmit the data, via relay satellites, to the Air Force Global Weather Center (AFG-WC) at the Offutt Air Force Base in Nebraska and the Fleet Numerical Oceanographic Center (FNOC) in Monterey, California.

3.4.1.2 The NOAA Satellite System (POES)

Satellites of the Polar-orbiting Operational Environmental Satellite (POES) system operate on sun-synchronous orbits. The NOAA satellite system adopts a double-satellite operation system. The local time of the orbit descending node of one of the satellites is in the morning, and that of the other is in the afternoon. Currently, the POES system satellites carry six kinds of payloads, which are shown in Table 3.25.

Table 3.25 Payloads of the POES system satellites

Satellite	Payloads
NOAA- K	AMSU-A, AMSU-B, ARGOS, ATOVS (HIRS/3 + AMSU + AVHRR/3), AVHRR/3, HIRS/3, NOAA Comms, S&R (NOAA)
NOAA- L	AMSU-A, AMSU-B, ARGOS, ATOVS (HIRS/3 + AMSU + AVHRR/3), AVHRR/3, HIRS/3, NOAA Comms, S&R (NOAA), SBUV/2, SEM (POES)
NOAA- M	AMSU-A, AMSU-B, ARGOS, AVHRR/3, HIRS/3, NOAA Comms, S&R (NOAA), SBUV/2, SEM (POES)
NOAA- N	AMSU-A, ARGOS, AVHRR/3, HIRS/4, MHS, NOAA Comms, S&R (NOAA), SBUV/2, SEM (POES)
NOAA- N'	A-DCS4, ARGOS, AVHRR/3, HIRS/4, LRIT, MHS, NOAA Comms, S&R (NOAA), SBUV/2, SEM (POES)

In these payloads, AVHRR/3 is used to detect clouds, and cloud-top, sea surface and land surface temperatures. Its channel characteristics are shown in Table 3.26.

HIRS/3 is used to sound the vertical profiles of atmospheric temperature and humidity on cloudless or partly cloudy days. With a quantization level of 13 bits, the instrument has 20 channels and a resolution of 17.4 km.

AMSU consists of AMSU-A and AMSU-B. AMSU can sound temperature and humidity on cloudy days, sound precipitation on the land and sea, recognize sea ice and determine its scope, and sound soil moisture to a certain degree.

SEM is used to measure solar protons, alpha particles, electron flux density, the energy spectrum, and the total particle energy distribution in the satellite orbit to study the satellite's physical environment in space, predict proton events, and ensure the safe operation of spacecraft working in orbit.

ERBS is used to observe incident solar shortwave radiation, solar shortwave radiation reflected to outer space, and longwave radiation transmitted from the Earth-atmosphere system. SBUV is used to measure the total amount and vertical distribution of ozone. The instrument detects the 160–400 nm band and measures two aspects: the ultraviolet backscatter of the atmosphere in the O_3 absorption band and the ultraviolet radiation of the Sun.

Table 3.26 Channel characteristics and applications of AVHRR/3

Channel	Wavelength (μm)	Resolution (km)	Typical application
1	0.58–0.68	1.09	Daytime cloud imaging
2	0.725–1.00	1.09	Ice and snow monitoring
3A	1.58–1.64	1.09	Aerosol, snow, and ice monitoring
3B	3.55–3.93	1.09	Fire and nighttime cloud imaging
4	10.30–11.30	1.09	Daytime and nighttime cloud imaging, land surface and sea surface temperature sensing
5	11.50–12.50	1.09	Daytime and nighttime cloud imaging, land surface and sea surface temperature sensing

Table 3.27 Payloads of third-generation GOES satellites in orbit

Satellite code	Payloads
GOES-12	DCS (NOAA), GOES Comms, Imager, LRIT, S&R (GOES), SEM (GOES), Sounder, SXI, WEFAX
GOES-13	A-DCS4, GOES Comms, Imager, LRIT, S&R (GOES), SEM (GOES), Sounder, SXI
GOES-14	A-DCS4, GOES Comms, Imager, LRIT, S&R (GOES), SEM (GOES), Sounder
GOES-15	A-DCS4, GOES Comms, Imager, LRIT, S&R (GOES), SEM (GOES), Sounder, SXI

3.4.1.3 The GOES Satellite System

The United States is now using the third generation of geostationary meteorological satellites. These satellites adopt a three-axis stabilization mode and a satellite-borne vertical sounder, and the imager can perform sounding separately at the same time. There are four main kinds of payloads. The orbital information and payloads of the Geostationary Operational Environmental Satellite (GOES) satellites currently in operation are shown in Table 3.27.

3.4.2 European Meteorological Observation Satellites

The European meteorological satellite program began in 1972. The initial goals of the program were to meet European countries' need for weather analysis and forecasting and meet the demand for global atmospheric monitoring and research in accordance with the WMO's World Weather Watch (WWW) program and the Global Atmospheric Research Program (GARP).

3.4.2.1 Typical Geostationary Meteorological Satellites of Europe

The European Organisation for the Exploitation of Meteorological Satellites (EUMETSAT) has launched ten Meteosat satellites since the first geostationary meteorological Meteosat satellite was launched in November 1977. The European geostationary meteorological satellites are the Meteosat series of satellites launched by EUMETSAT; Meteosat-7 belongs to the first generation (Fig. 3.24) and Meteosat-8, Meteosat-9 and Meteosat-10 belong to the second generation.

The main instrument installed on the first generation of Meteosat operational satellites is a three-channel imager, MVIRI. The parameters of each channel are listed in Table 3.28. The satellites' main tasks are to (1) provide 48 full-disk images of Earth daily; (2) transmit near-real time digital and analog images to primary data user stations and secondary data user stations; (3) relay image data transmitted from

Fig. 3.24 First-generation Meteosat system

Table 3.28 Features of first-generation Meteosat operational satellites

Channel	Spectrum (μm)	Pixel × scan line
Visible (VIS)	0.5–0.9	5000 × 5000
Infrared (IR)	10.5–12.5	2500 × 2500
Water vapor (WV)	5.7–7.1	2500 × 2500

other meteorological satellites; (4) collect data transmitted from the data acquisition platform; (5) send meteorological products to users; and (6) perform meteorological data distribution (MDD), which is mainly intended to improve the transmission of African meteorological data.

The second-generation Meteosat satellites entered Phase A (system design phase) before 1993 and entered Phase B (sample satellite development phase) soon after. Phase C was developed as the launch and implementation phase, and Phase D was the postlaunch application and improvement phase.

MSG is a spin-stabilized satellite (Fig. 3.25), similar to the first generation of meteorological satellites. Its design was improved in many aspects. For instance, the satellite-borne radiometer SEVIRI has much higher performance, the spectral channels were increased from three to twelve, the resolution was greatly improved (1 km in the wideband high-resolution visible light channel), and the scanning time was halved from thirty minutes to fifteen minutes. The data transmission system was also improved, making data transmission and broadcast much faster (3.2 Mbps and 1 Mbps, respectively).

Fig. 3.25 MSG satellite

3.4.2.2 Polar-Orbiting Meteorological Satellite System

The European Union's polar-orbiting meteorological satellite system, MetOp, and EUMETSAT teams are working closely together to develop a European polar-orbiting meteorological satellite system and launch the MetOp series of satellites which, starting in 2002, began replacing older meteorological satellites (TIROS series) launched earlier by NOAA. Satellites owned and operated by EUMETSAT will be part of an American-European three-satellite operating system, in which one US satellite will appear at dawn, MetOp will appear in the morning and another US satellite will appear in the early afternoon.

MetOp is being designed to carry instruments provided by the ESA, EUMETSAT, NOAA, and CNES. These satellites have a larger carrying capacity, improved payload, and better performance than the NOAA system. The MetOp series consists of three satellites; the first, MetOp-A (Fig. 3.26), was launched on October 19, 2006, with a design life of five years and the second, MetOp-B (Fig. 3.27), was launched on September 1, 2012.

The EUMETSAT polar-orbiting satellite system is an integral part of the global observing system (GOS) that is designed to provide long-term global observation

Fig. 3.26 MetOp-A

Fig. 3.27 MetOp-B

data in conjunction with NOAA satellites. The operational instruments on board the EUMETSAT polar-orbiting system are designed to be the same as those on board NOAA satellites to ensure the consistency of observation data. The first one or two satellites are large-capacity, nonoperational polar-orbiting platforms (EPOP/POEM), and subsequent satellites are smaller MetOp satellites.

3.4.3 *China's Meteorological Observation Satellites*

China's polar-orbiting meteorological satellites (FY-1 and FY-3 satellite series) are also referred to as sun-synchronous orbiting meteorological satellites, those whose orbital plane is usually 98°–99° from the equatorial plane and whose orbit crosses the north and south poles. Geostationary meteorological satellites (FY-2 satellite series) move at the same speed as Earth's rotation at an altitude of 36,000 km above the equator. Information on the FY satellite series is shown in Fig. 3.28 (National Satellite Meteorological Center, China Meteorological Administration 2013a).

3.4.3.1 Polar-Orbiting Satellites

(1) *FY-1A/1B*

FY-1A was launched on September 7, 1988, as an experimental meteorological satellite. Although it only worked in orbit for 39 days due to a control system failure, the successful launch of FY-1A was considered a milestone in China's development of meteorological satellites. The satellite was equipped with an infrared and visible light scanning radiometer, a data collection system, a space environment detector, and other instruments. Technical parameters of the multispectral infrared and visible light scanning radiometer are shown in Table 3.29 (National Satellite Meteorological Center, China Meteorological Administration 2013b).

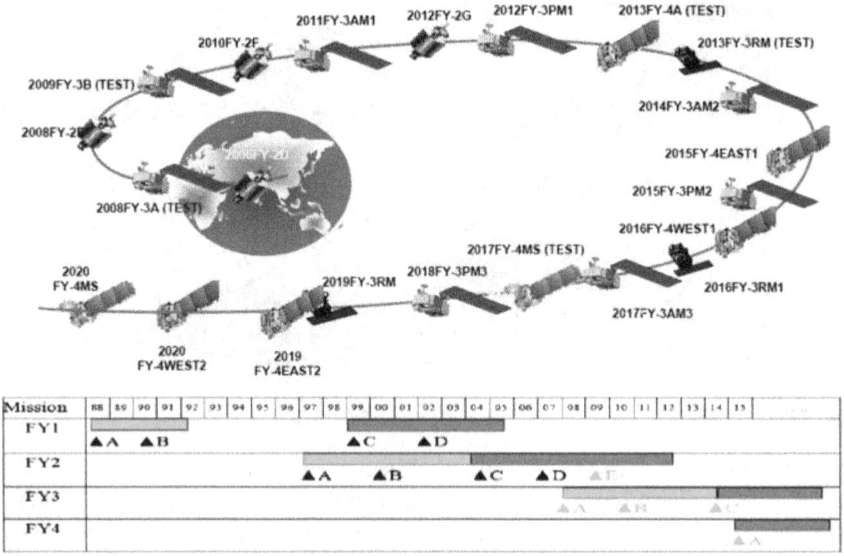

Fig. 3.28 FY satellite series (CMA)

Table 3.29 Technical parameters of FY-1A's visible and infrared scanning radiometer

Component	Parameter
Sensor	Multispectral infrared and visible light scanning radiometer
Tasks	To acquire day-and-night visible light, infrared cloud imagery, snow and ice cover, vegetation, ocean color, sea surface temperature, etc.
Scan rate	6 scanning lines/second
Earth-scanning angle (°)	±55.4
Nadir ground resolution (km)	1.1
Data quantization level (bit)	10
Calibration accuracy	Visible and near infrared channels 10% (reflectance); infrared channels 1 K (300 K)
Wavelength (μm)	0.58–0.68, 0.725–1.1, 0.48–0.53, 0.53–0.58, and 10.5–12.5
Data transmission	For high-resolution picture transmission (HRPT), the bit rate is 0.6654 Mbps and the operating frequency is 1,670–1,710 MHz. In low-resolution image transmission (APT), delay picture transmission (DPT), high-resolution picture transmission (APT) and DPT are analog signals

The FY-1B satellite was successfully launched on September 3, 1990. As China's second experimental meteorological satellite, FY-1B was an improvement over FY-1A. Compared with FY-1A, FY-1B's attitude control system was improved and its visible cloud images were clearer. The performance of the satellite's sensors and the main functions of the satellite were similar to those of the United States' third-generation polar-orbit meteorological satellites. The satellite's performance was at a level similar to that of commercial applications, its visible channel image quality was high, and its signal-to-noise ratio was above the design requirement. However, the satellite's system lacked reliability.

(2) *FY-1C*

FY-1C was successfully launched from the Taiyuan Satellite Launch Centre on May 10, 1999. Compared with FY-1A/B, the FY-1C satellite had significantly improved performance, with increased detection channels and accuracy. Its design life was two years. A series of technical measures were taken that led to improvements in the product quality, adaptability to space environments, and system reliability. FY-1C functioned stably in orbit until June 24, 2004, when the reception of FY-1C cloud images ceased.

The satellite was equipped with a space particle composition detector and a multichannel visible infrared scanning radiometer (MVISR). The number of MVISR channels for FY-1C was increased from five (FY-1A) to ten, and included four visible light channels, one shortwave infrared channel, and two longwave infrared channels. Table 3.30 lists the wavelength and use of each channel. The field of view was 1.2 microradians, the nadir resolution was 1.1 km, and the scanning speed was six scan lines per second, with each line containing 20,480 pixel points. The calibration

Table 3.30 Technical parameters of FY-1C's multispectral infrared and visible scanning radiometer

Channel no.	Wavelength (μm)	Main purpose
1	0.58–0.68	Daytime clouds, ice, snow, vegetation
2	0.84–0.89	Daytime clouds, vegetation, water
3	3.55–3.93	Heat sources, nighttime clouds
4	10.3–11.3	Sea surface temperature, day/nighttime clouds
5	11.5–12.5	Sea surface temperature, day/nighttime clouds
6	1.58–1.64	Soil moisture, ice and snow recognition
7	0.43–0.48	Ocean color
8	0.48–0.53	Ocean color
9	0.53–0.58	Ocean color
10	0.90–0.965	Water vapor

accuracy of the visible and near infrared channels reached 10%, and the infrared radiometric calibration accuracy reached 1 K, as technically required. The spatial resolution of the HRPT and GDPT images was greater than 1.1 km and 4 km, respectively. The Chinese high-resolution picture transmission (CHRPT) had a frequency of 1,700 MHz, a bit rate of 1.3308 Mbps, and real-time reception from anywhere in the world. The delay picture transmission (DPT) had a frequency of 1,708 MHz and a bit rate of 1.3308 Mbps and was divided into two types: GDPT and LDPT (National Satellite Meteorological Center, China Meteorological Administration 2013b).

(3) *FY-1D*

Design of the FY-1D flight model began in 2000 based on FY-1C technology and previous experience. Fourteen technical improvements were made that led to improved stability. The 950 kg satellite was launched from the Taiyuan Satellite Launch Center on May 15, 2002, using an LM-4B rocket. FY-1D functioned normally for ten years, exceeding its design life and completing all tasks. It is no longer in operation.

FY-1D's main onboard sensor was a multichannel visible infrared scanning radiometer (MVISR), whose main technical parameters are listed in Table 3.31. Data were transmitted using two methods: HRPT and DPT. The HRPT's bit rate was 1.3308 Mbps, and the carrier frequency was 1,700.4 MHz. The DPT's bit rate was 1.3308 Mbps, and the carrier frequency was 1,708.46 MHz. Global meteorological data could be acquired through four channels (Channels 1, 2, 4 and 5), with a spatial resolution of 3.3 km (National Satellite Meteorological Center, China Meteorological Administration 2013b).

(4) *FY-3A*

The FY-3 satellites (FY-3) were China's second-generation polar-orbiting meteorological satellites used for weather forecasting, climate prediction, and environmental monitoring. The FY-3 series comprised two satellites: FY-3A and FY-3B. The satellites were used to conduct 3D atmospheric detection, greatly improved China's ability

Table 3.31 Technical parameters of FY-1D's multispectral infrared and visible scanning radiometer

Component	Parameter
Sensor	Multispectral infrared and visible light scanning radiometer
Tasks	To acquire day-and-night visible light, infrared cloud imagery, snow and ice cover, vegetation, ocean color, sea surface temperature, etc.
Scan rate	6 scanning lines/second
Earth-scanning angle (°)	±55.4
Nadir ground resolution (km)	1.1
Data quantization level (bit)	10
Calibration accuracy	Visible and near infrared channels 10% (reflectance); infrared channels 1 K (300 K)
Wavelength (μm)	0.58–0.68, 0.84–0.89, 3.55–3.93, 10.3–11.3, 11.5–12.5, 1.58–1.64, 0.43–0.46, 0.48–0.53, 0.53–0.58, and 0.900–0.965
Data transmission	For high-resolution picture transmission (HRPT), the bit rate is 0.6654 Mbps and the operating frequency is 1,670–1,710 MHz. In low-resolution image transmission (APT), delay picture transmission (DPT), high-resolution picture transmission (APT) and DPT are analog signals

to acquire global information and further enhanced its cloud area and surface feature remote sensing capabilities. These features enabled the country to obtain global, all-weather, three-dimensional, quantitative, multispectral data on atmospheric, land surface, and sea surface characteristics.

FY-3A was the first FY-3 meteorological satellite launched using an LM-4C rocket from the Taiyuan Satellite Launch Center at 11:02 on May 27, 2008. Although it was developed based on the FY-1 meteorological satellites, FY-3A was substantially superior in both technology and function. The satellite was capable of three-dimensional atmospheric detection, greatly improving the capability for global information acquisition and cloud area and surface feature remote sensing.

(5) *FY-3B*

FY-3B is the second satellite in the FY-3 meteorological satellite series. It was launched from the Taiyuan Satellite Launch Center in the early morning of November 5, 2010, using an LM-4C rocket. FY-3B is China's first afternoon-orbit meteorological satellite, making it the first polar-orbiting meteorological satellite to conduct observations at this time. FY-3B is useful for accurate monitoring and numerical forecasting of rainstorms in southern China that usually occur in the afternoon. Working in conjunction, FY-3B and FY-3A increased the global scan frequency from twice a day to four times a day. Thus, China's ability to monitor disastrous weather events such as typhoons and thunderstorms was enhanced markedly. The satellite had a design life of three years but is still operating in orbit.

FY-3B is equipped with eleven advanced remote sensing instruments and 99 spectral detection channels, five of which have a resolution of 250 m. FY-3B is similar to FY-3A in terms of the satellite platform, payload configuration, and main performance parameters. However, as the first next-generation, polar-orbiting meteorological satellite, FY-3A showed weak operation of some onboard instruments. FY-3B was developed by meteorological satellite experts based on their experience acquired from the development of the FY-3A satellite. As a result, FY-3B demonstrated improved performance for the infrared spectrometer, microwave radiation imager, and solar backscatter ultraviolet sounder.

(6) *FY-3C/3D*

FY-3C is a sun-synchronous orbit satellite launched on September 23, 2013 by the carrier rocket Chinese Long March 4C from the Taiyuan Satellite Launch Center in Shanxi province. The FY-3C orbital satellite joins its predecessors FY-3A and FY-3B. It replaced FY-3A to operate, after undergoing tests, in a morning orbit with FY-2B, which is in an afternoon orbit, to provide temporal resolution of global observation data of up to six hours.

The FY-3C missions primarily include Earth surface imaging and atmospheric sounding, and its observational data will be used in weather forecasting, and in monitoring of natural disasters and ecological and environmental factors. Compared with FY-3A and FY-3B, the payload on board FY-3C features 12 sensing instruments, including a visible infrared radiometer, a microwave scanning radiometer, a microwave temperature sounder (MWTS), a microwave humidity sounder (MWHS), a microwave imager, and a medium resolution imaging spectrometer. It also includes a UV-O-zone sounder, a total O-zone UV detector, a solar radiation and Earth radiation detector, space environmental monitoring suits, and GNSS occultation detectors.

FY-3D was launched on November 14, 2017 as China's fourth second-generation polar-orbiting meteorological satellite and will replace the orbiting FY-3B satellite. The satellite is designed to provide weather forecasts in medium- and long-range numerical weather prediction (NWP) models, enabling high-impact weather forecasting up to a week in advance, and alleviate the impacts of natural disasters on the economy and society and improve livelihood.

Equipped with greenhouse gas probing capacity, FY-3D was also developed to help tackle climate change, in addition to serving ecological, civilization, and construction needs and the 'Belt and Road' initiative. FY-3D features ten instruments, including a microwave temperature sounder (MWTS), a microwave humidity sounder (MHTS), a microwave radiation imager (MWRI), a space environment monitor (SEM), and a global navigation satellite system occultation sounder (GNOS).

3.4.3.2 Geostationary Orbit Satellites

(1) *FY-2A/2B*

The FY-2A satellite was the first experimental satellite in China's first-generation geostationary meteorological satellite series, FY-2, and was launched on June 10, 1997. FY-2A had a three-channel scanning radiometer and a design life of three years at a stable spinning altitude. The satellite began to have issues after working for three months and then worked intermittently, only operating for six to eight hours each day. Ultimately, FY-2A failed to meet the requirements for commercial meteorological services.

The main payload of FY-2A was a visible and infrared spin-scan radiometer (VISSR), whose technical parameters are shown in Table 3.32 (National Satellite Meteorological Center, China Meteorological Administration 2013b).

The FY-2B satellite was the second experimental satellite in China's first-generation geostationary meteorological satellite FY-2 series. FY-2B was launched on June 25, 2000 from the Xichang Satellite Launch Center using an LM-3 rocket. The first original cloud image was received on July 6. FY-2B only had a three-channel scanning radiometer and a design life of three years in a stable spinning altitude. It functioned in orbit for less than eight months before a problem occurred with one of the components on board the satellite; from then onward, the signals it sent back were too weak to receive. Ultimately, FY-2B failed to meet the requirements for commercial meteorological services. However, FY-2B's operation provided valuable experience for the development of subsequent FY-2 meteorological satellites.

The technical parameters of the FY-2B and FY-2A satellites were identical. The cloud images sent from FY-2B played an important role in monitoring typhoons and marine weather, forecasting rainstorms, preventing floods, analyzing the weather system above the Qinghai-Tibetan Plateau, providing meteorological support for aviation, and predicting climate change.

(2) *FY-2C/2D/2E/2F/2G/2H*

FY-2C was the first commercial-use satellite in the FY-2 meteorological satellite series. After a successful launch on October 19, 2004, FY-2C was positioned at an altitude of 36,000 km above the equator at 105° east longitude on October 24.

Table 3.32 Technical parameters of FY-2A's visible and infrared light spin-scan radiometer (VISSR)

Channel	Visible light	Infrared	Water vapor
Wavelength (μm)	0.55–1.05	10.5–12.5	6.2–7.6
Resolution (km)	1.25	5	5
Field of view (μrad)	35	140	140
Scan line	2,500 × 4	2,500	2,500
Detector	Si-photo-diode	HgCdTe	HgCdTe
Noise resolution	S/N = 6.5 (Albedo = 2.5%) S/N = 43 (Albedo = 95%)	NEDT = 0.5–0.65 k (300°K)	NEDT = 1 K (300°K)
Quantitative byte (bit)	6	8	8
Scanning step	140 μrad (N-S scanning)		

FY-2C occupied FY-2B's former position to monitor weather conditions in the Asia Pacific Region. Four days after it was positioned, adjustments were made to the ground application system to technically coordinate it with the satellite. The satellite's service monitoring, data transmission, and forwarding channels were opened, and the scanning radiometer was switched on. FY-2C could observe changes in sea surface temperature, and one of its channels was designed for measuring 3.5–4 μm light waves to observe high-temperature heat sources on the ground. It was possible to use spectral channels to observe ground heat sources to promptly discover forest fires in remote and desolate places, monitor their situation, and predict their development trends.

FY-2D was the fourth satellite in the FY-2 meteorological satellite series. FY-2D was also the country's second application-oriented geostationary-orbiting meteorological satellite. It was launched using an LM-3A rocket at 08:53 on December 8, 2006. After 1,421 s of flying, it successfully separated from the rocket, entering into a large elliptical transfer orbit with a perigee altitude of 202 km, apogee altitude of 36,525 km, and inclination of 24.97°. At 01:24 on December 9, the apogee engine was ignited for orbital transfer, and secondary separation was successfully completed. After four batches of orbit trimming, the satellite was positioned at an altitude of 36,000 km above the equator at 86.5° E longitude at 17:00 on December 13. It is currently no longer in operation.

On December 23, 2008, and January 13, 2012, China's third and fourth service-oriented geostationary meteorological satellites, FY-2E and FY-2F, respectively, were launched from the Xichang Satellite Launch Center using LM-3A rockets. The two satellites were of great significance for the continuous and stable operation of China's geostationary meteorological satellite observation services. FY-2F boasted flexible capability for scanning specific regions with a high temporal resolution and could monitor disastrous weather conditions such as typhoons and severe convections. FY-2F played an important role in China's meteorological disaster monitoring, early warning, prevention, and mitigation. The space environment monitor continuously monitored solar X-rays and the flow of high-energy protons, electrons, and heavy particles, and the data were used for space weather monitoring, forecasting, and early warning services.

The geostationary meteorological satellites FY-2C, FY-2D, FY-2E, and FY-2F working in orbit formed a "double-satellite observation with mutual backup" service pattern. These satellites helped modernize China's comprehensive meteorological observation system. During flood season, the double-satellite observation mode allowed for spinning the satellite, enabling it to provide a cloud picture every fifteen minutes. This intensified observation mode played a key role in monitoring disastrous weather systems such as typhoons, rainstorms, thunderstorms, and small- and medium-scale local convective systems. The FY-2 meteorological satellite series played a crucial role in combating heavy rain, freezing snow, and other extreme weather events. The satellites also provided assistance in the Wenchuan earthquake relief operations and in providing meteorological services for the Beijing Olympics and Paralympics.

The FY-2G satellite was launched on December 31, 2014 from the Xichang Satellite Launch Center. Based on the technology of FY-2 F satellite, the FY-2G satellite was improved by reducing infrared stray radiation, uplifting the observation frequency for the blackbody, and improving the telemetry resolution of optical components. These improvements increase the retrieval accuracy of FY-2G satellite quantitative products and enhance the quantitative application of satellite data products.

FY-2H was launched on June 5, 2018. It is positioned over the Indian Ocean and has realized the sustained observation of one-third of the Earth's territories from Oceania to central Africa. It can provide favorable observation perspectives and custom-made high-frequency subregional observation for countries and regions such as western Asia, central Asia, Africa, and Europe. Equipped with a scanning radiometer and a space environment monitor, FY-2H can supply data for dozens of remote sensing products such as cloud images, clear sky atmospheric radiation, sand and dust, and cloud motion wind (CMW) for weather prediction, disaster warning, and environmental monitoring, enriching the data sources for global NWP models.

The main payload of FY-2C/2D/2E/2F was a visible and infrared spin-scan radiometer (VISSR), whose technical parameters are shown in Table 3.33 (National Satellite Meteorological Center, China Meteorological Administration 2013b).

(3) *FY-4A*

FY-4A was launched on December 11, 2016, as the first Chinese second-generation geostationary meteorological satellite. FY-4A is China's first quantitative remote sensing satellite with a three-axis stabilization structure on a geostationary orbit. Four new instruments are on board the latest independently developed weather satellite, namely, an advanced geosynchronous radiation imager (AGRI), a geosynchronous interferometric infrared sounder (GIIRS), a lightning mapping imager (LMI) and a space environment package (SEP).

FY-4A is the first satellite in China that can capture lightning. The onboard Lightning Mapping Imager enables this function. It is the first geostationary optical remote sensing instrument in China and has filled the gap in terms of lightning observation and satellite-borne detection. FY-4A can detect lightning over China and neighboring areas and take 500 lightning pictures per second. By real-time and consecutive observation of lightning, it can aid in observation and tracking of severe convective weather and provide early warning for lightning disasters.

Table 3.33 Technical parameters of the radiometer on board FY-2C/2D/2E/2F

Channel	Waveband (μm)	Resolution (km)
Visible light	0.55–0.90	1.25
Infrared 1	10.3–11.3	5
Infrared 2	11.5–12.5	5
Infrared 3	6.3–7.6	5
Infrared 4	3.5–4.0	5

3.4.4 Other Meteorological Observation Satellites

3.4.4.1 Japan's Satellites

Since Japan launched its first geostationary meteorological satellite, GMS-1, in 1977, it has put five geostationary meteorological satellites into orbit. The GMS-4 satellite is positioned at 140°E above the equator and is equipped with visible and infrared scanning radiometers that observe a fourth of Earth to monitor cloud distribution, height and dynamics. The satellite can obtain information about winds below and above clouds, and detect sea surface temperature distribution.

Similar to other GMS satellites, GMS-5 is a spin-stabilized satellite. Its total mass is 756 kg, the design life is five years, and the main onboard instrument is a visible and infrared light spin-scan radiometer (VISSR). The VISSR was significantly improved by building upon the radiometer on board GMS-4. One 6.5–7 μm WV channel was added to observe water vapor radiation in the middle layer of the troposphere. The original 10.5–12.5 μm infrared window area was split into a 10.5–11.5 μm channel and an 11.5–12.5 μm channel to observe radiation from Earth's surface and the atmosphere. The nadir spatial resolution of GMS-5 is 1.25 km for the visible light channel and 5 km for the WV channel. The main parameters of the VISSR on board GMS-5 is listed in Table 3.34.

After GMS-5 was launched, Japan suspended the development of single-function meteorological satellite systems. The Japan Meteorological Agency and Japan Civil Aviation Administration jointly developed a new large, multifunctional, integrated satellite system called MTSAT. MTSAT-1, the first satellite of this system, was scheduled to be launched on November 15, 1999. However, the launch was unsuccessful due a fault with the rocket and both the satellite and rocket were destroyed. Japan manufactured another MTSAT satellite named MTSAT-1R (Kim et al. 2011). The satellite was not launched until February 26, 2005 due to the time required to remove the fault and improve the rocket. The satellite began to broadcast images two to three months after launch. It was followed by MTSAT-2, which was launched on December 26, 2006 (Fig. 3.29). The MTSAT satellites are equipped with VISSR, cloud image broadcasting, DCS, aviation communication, and other subsystems mainly used for meteorological exploration and aviation communication and are the largest geostationary satellites with meteorological sounding functions (Crespi et al. 2012).

Table 3.34 VISSR parameters of the GMS-5 satellite (Huang et al. 2004)

Channel	Wavelength (μm)	Quantization level (bit)	Spatial resolution (nadir) (km)
Visible light	0.55–0.90	8	1.25
Water Vapor (WV)	6.5–7.0	8	5
Infrared window area	10.5–11.5	8	5
Infrared window area	11.5–12.5	8	5

Fig. 3.29 The MTSAT-2 satellite

3.4.4.2 India's Satellites

INSAT is a multiagent multitarget satellite system and is one of the largest satellite systems in Asia. The INSAT satellite system has played an increasingly important role in the Indian aerospace industry with the continuous development and improvement of the INSAT-1, INSAT-2, and INSAT-3 series of satellites.

INSAT provides services such as domestic long-distance communication, meteorological and Earth observation data relay, augmented television receiver national direct satellite broadcasting, TV education, rural communications, meteorology, and disaster alarms.

The first-generation INSAT satellites, the INSAT-1 series, were manufactured by Ford Motor Co. in the United States and comprised four satellites: INSAT-1A, INSAT-1B, INSAT-1G, and INSAT-1D. The second-generation INSAT satellites, INSAT-2, were independently developed by India to meet the needs of the 1990s. The INSAT-2 series consisted of five satellites: INSAT-2A, INSAT-2B, INSAT-2C, INSAT-2D, and INSAT-2E. In addition to normal C-band transponders, the INSAT-2 satellites also adopted the high-frequency section of the C-band, or the extended C-band. The third-generation INSAT satellites, INSAT-3, were also made by the Indian Space Research Organization (IRSO) and comprise five satellites: INSAT-3A, INSAT-3B, INSAT-3C, INSAT-3DR, and INSAT-3DS.

INSAT-3A is a multipurpose satellite launched on April 10, 2003 using an Ariane rocket. The satellite is fixed at 93.50° E and has the following payloads (Fig. 3.30). Of the twelve C-band transponders, nine provide coverage that extends from the Middle East to southeast Asia with an EIRP of 38 dBW, and three provide coverage of India with an EIRP of 36 dBW. The six extended C-band transponders provide

Fig. 3.30 The INSAT-3A satellite

Indian coverage, with an EIRP of 36 dBW. The six Ku-band transponders also provide coverage of India, with an EIRP of 48 dBW. The one very high-resolution radiometer (VHRR) can perform imaging in the visible light channel (0.55–0.75 μm), thermal infrared channel (10.5–12.5 μm), and water vapor channel (5.7–7.1 μm) with ground resolutions of 2 × 2 km, 8 × 8 km, and 8 × 8 km, respectively. The CCD camera has a ground resolution of 1 × 1 km in the visible (0.63–0.69 μm), near infrared (0.77–0.86 μm), and SWIR (1.55–1.70 μm) channels.

3.4.4.3 Russia's Satellites

(1) *Russia's polar-orbiting meteorological satellites*

The "Meteor" series of polar-orbiting meteorological satellites was developed by the Soviet Union/Russian Federation and has gone through four generations. Most of the previous three generations of satellites do not function in sun-synchronous orbit. However, the fourth-generation of satellites is known to work in a sun-synchronous orbit.

As early as 1962–1969, the Soviet Union had launched more than 20 COSMOS satellites for meteorological observation. In March 1969, it launched its first-generation polar-orbiting meteorological satellite: Meteor-1. The first generation consisted of 31 satellites (Meteor-1-31) launched from 1969 to 1981, most of which had an orbital inclination of 81.2°. The second generation (Meteor-2) comprised 24 satellites launched after 1975. In most cases, two or three satellites were simultaneously operating on orbit, with an orbital inclination of 82.0° and orbital altitude of 950 km. The third-generation (Meteor-3) polar-orbiting meteorological satellites were launched in 1984. The third generation was composed of eight satellites, which had an orbital inclination of 82° and orbital altitude of 1,200 km.

Meteor-3 M N1, the first satellite of the fourth generation of Russian meteorological satellites, (the Meteor-3 M series) was launched on December 10, 2001 (Fig. 3.31).

Major changes in the Meteor-3 M series of satellites include: 99.6° orbital inclination, 1,024 km sun-synchronous orbit, and a broadcast data format that is compatible with NOAA's high-resolution picture transmission (HRPT).

(2) *Russia's geostationary orbit meteorological satellites*

Russia's first geostationary orbit meteorological satellite (GOMS) was successfully launched in November 1994. It is a three-axis stabilized satellite positioned at 76°E. A problem occurred with the attitude control after launch, but the satellite resumed working after some remedial measures were taken. Unfortunately, its scanning radiometer's visible light channel has been unable to acquire any images due to an optical design error; thus, the satellite can only capture infrared images.

On January 20, 2011, Russia launched the geostationary hydrological and meteorological satellite Elektro-L from the Baikonur Launch Center in Kazakhstan (Fig. 3.32). Fixed at a position 36,000 km above Earth, the satellite is used to monitor climate change in Russia's Asian region. The visible light and infrared photographic devices installed on the satellite can capture 1-km and 4-km resolution ground images, respectively. Under normal circumstances, the satellite takes a photo once every 30 min. The shooting frequency can be increased to once every 10–15 min in the event of a natural disaster. The satellite is also responsible for forwarding and exchanging weather information as well as receiving and forwarding signals from the international search and rescue satellite COSPAS-SARSAT. GOMS has a life span of ten years and the data distribution mode is HRPT/LRPT. Its mission is to observe

Fig. 3.31 The Meteor-3 M satellite

Fig. 3.32 The Electro-L satellite

Earth's surface and atmosphere, perform solar-geophysical measurement, and support the data collection system and COSPAS-SARSAT services. The satellite's main payload is an optical imaging radiometer, MSU-GS, which provides imaging data in three VNIR channels and seven infrared channels. Its nadir spatial resolution (sampling distance) is approximately 1 km (for visible light) and 4 km (for VNIR and infrared), with a new Earth image provided once every 30 min.

3.5 Trends in Remote Sensing for Digital Earth

Looking back on the past five decades of spaceborne remote sensing, every step along the way has been based on the national backgrounds and political and economic conditions of each country. During this period of development, the purpose of Earth observation shifted from single-field surveying toward serving the demands of the overall development of human society (Guo 2014). Since entering the period of globalization, remote sensing technologies have developed into a complete system (Guo et al. 2013), which will provide more abundant data for Digital Earth.

Countries and regions with leading Earth observation technologies, such as the United States and Europe, have formulated Earth observation plans for long-term development. In 2013, the United States and European organizations were expected to launch 34 Earth observation satellites, and India and China planned to launch 25 and 26 satellites, respectively. Russia, Japan, and Canada also had plans for over ten launch missions (Fig. 3.33). Russia will remain a major contributor to satellite launches in Europe, but European organizations will launch significantly more, and there will be a greater emphasis on cooperation and coordination between European

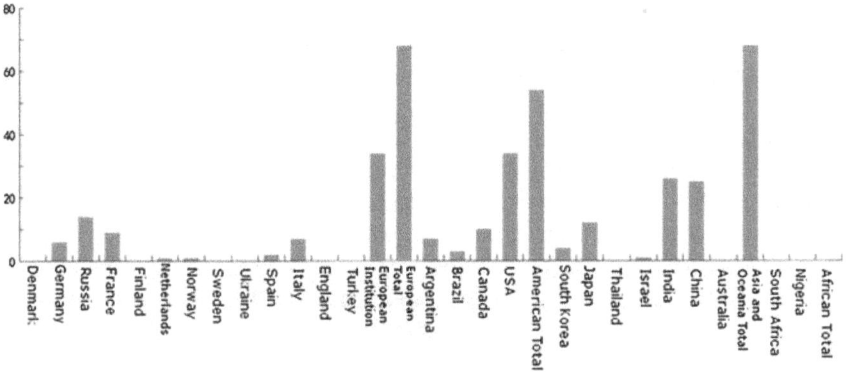

Fig. 3.33 Global launch plans for Earth observation satellites by 2035

countries. In America, the United States will remain a leading force, and Canada will occupy a secondary role. In Asia, the existing trend will continue, with China, India, Japan, and South Korea continuing to be major contributors. Currently, no African countries have plans to launch new satellites.

All of the aforementioned satellite programs have clearly defined services. For example, the United States' Earth observation program for 2016–2020 focuses on measuring global ozone conditions and other relevant gases (GACM program), atmospheric pollution monitoring (3D-Winds), geological disasters (LIST), weather forecasts (PATH), and water resource utilization (GRACE-II/SCLP) (Neeck et al. 2008). The European GMES program covers the six service fields of land, ocean, emergency management, security, atmosphere, and climate change (Veefkind et al. 2012). In addition, Russia, Japan, India, and some other countries have issued strategic plans for Earth observation, forming systems with their own characteristics. The Russian Federal Space Agency (Roscosmos) intends to form a satellite system consisting of geostationary meteorological satellites (Elektro series), polar-orbiting meteorological satellites (METEOR series), and resource/environment satellites (KANOPUS-V and Resurs-P series) by 2020. The Japan Aerospace Exploration Agency (JAXA) proposed the GOSAT program for greenhouse gas monitoring and the GCOM program for global change monitoring in addition to its ongoing efforts to build the ALOS program of high spatial resolution satellites carrying L-band SAR and hyperspectral sensors. Additionally, JAXA has plans to continue with its navigation experiment satellite program (QZS). The Indian Space Research Organization (ISRO) and the Indian National Remote Sensing Agency (NRSA) aim to improve the spatial resolution of the Resourcesat series and develop SAR-carrying satellites and environment satellites (Environment Sat) of their own (RISAT series).

In addition, some companies such as DigitalGlobe are planning to deploy new high-resolution satellites and trying to enter the microsatellite field. The planned satellites have also been extended from optical to meteorological and radar satellites. However, at present, there are few companies in the commercial satellite market; for example, DigitalGlobe provides high-resolution optical images, and the European

Airbus Defence and Space division can provide high-resolution optical and radar data. China is also planning a series of microsatellites for commercial service. Shenzhen-1 is its first microsatellite constellation and will realize 0.5 m resolution with a revisit period of less than 1 day. Furthermore, the Zhuhai-1, Beijing-1 and Beijing-2 microsatellites will be launched successively and networked. These commercial microsatellites aim to provide real time information for Digital Earth.

Future Earth satellite observation programs will focus on program continuity, development potential, and the capacity for comprehensive and coordinated applications. Therefore, long-term observation programs will be proposed and the development of aircraft-carried and satellite-carried sensors will continue with improved coordination. Relevant Earth observation programs will emphasize the coordinated use of Earth observation platforms and data to better meet the requirements of various fields that may benefit from observation efforts, as well as the nuanced strategic goals of countries and regions (Guo 2018).

References

Aguilar MA, Saldaña MM, Aguilar FJ (2013) GeoEye-1 and WorldView-2 pan-sharpened imagery for object-based classification in urban environments. Int J Remote Sens 34(7):2583–2606

Ali AM, Darvishzadeh R, Skidmore AK (2017) Retrieval of specific leaf area from landsat-8 surface reflectance data using statistical and physical models. IEEE J Sel Top Appl Earth Obs Remote Sens 10(8):3529–3536

Badura T, Sakulin C, Gruber C et al (2006) Derivation of the CHAMP-only global gravity field model TUG-CHAMP04 applying the energy integral approach. Stud Geophys Geod 50(1):59–74

Bouman J, Rispens S, Gruber T et al (2009) Preprocessing of gravity gradients at the GOCE high-level processing facility. J Geod 83(7):659–678

Chander S, Chauhan P, Ajai P (2012) Variability of altimetric range correction parameters over Indian tropical region using JASON-1 & JASON-2 radar altimeters. J Indian Soc Remote Sens 40(3):341–356

China Academy of Space Technology (2004) China-Brazil Earth Resource Satellite (CBERS). http://www.cast.cn. Accessed 10 Dec 2018

China Center for Resources Satellite Date and Application (2012) Landsats. http://www.cresda.com/. Accessed 10 Dec 2018

Crespi M, Fratarcangeli F, Giannone F et al (2012) A new rigorous model for high-resolution satellite imagery orientation: application to EROS A and QuickBird. Int J Remote Sens 33(8):2321–2354

Dash P, Walker N, Mishra D et al (2012) Atmospheric correction and vicarious calibration of Oceansat-1 Ocean Color Monitor (OCM) data in coastal case 2 waters. Remote Sens 4(6):1716–1740

DeVisser MH (2013) Exploration of sensor comparability: a case study of composite MODIS aqua and terra data. Remote Sens Lett 4(6):599–608

Dibarboure G, Renaudie C, Pujol MI et al (2011) A demonstration of the potential of Cryosat-2 to contribute to mesoscale observation. Adv Space Res 50(8):1046–1061

Ding J (2013) GF-1 to be launched. China Science Daily, fourth edition

Ditmar P (2004) A technique for modeling the Earth's gravity field on the basis of satellite accelerations. J Geod 78(1):12–33

Dong FX (2012) The current situation and future of foreign marine satellites. Space Explor 12:24–27

Floberghagen R, Fehringer M, Lamarre D et al (2011) Mission design, operation and exploitation of the gravity field and steady-state ocean circulation explorer mission. J Geod 85(11):749–758

GFZ (2018) The CHAMP Satellite. https://www.gfz-potsdam.de/en/section/geomagnetism/infrastructure/champ/satellite-systems/. Accessed 25 Jan. 2019
Gohil BS, Sikhakolli R, Gangwar RK (2013) Development of geophysical model functions for Oceansat-2 scatterometer. IEEE Geosci Remote Sens Lett 10(2):377–380
Gooding RH, Wagner CA, Klokočník J et al (2007) CHAMP and GRACE resonances, and the gravity field of the Earth. Adv Space Res 39(10):1604–1611
Goward SN, Chander G, Pagnutti M et al (2012) Complementarity of ResourceSat-1 AWiFS and Landsat TM/ETM+ sensors. Remote Sens Environ 123:41–56
Guo H (2012) China's Earth observing satellites for building a digital Earth. Int J Digit Earth 5(3):185–188
Guo H (2014) Special section guest editorial: Earth observation for global environmental change. J Appl Remote Sens 8(1):1–3
Guo H (2018) Steps to the digital silk road. Nature 554(7690):25–27
Guo H, Ding Y, Liu G et al (2014a) Conceptual study of lunar-based SAR for global change monitoring. Sci China Earth Sci 57(8):1771–1779
Guo H, Dou C, Zhang X et al (2016) Earth observation from the manned low Earth orbit platforms. ISPRS J Photogramm Remote Sens 115:103–118
Guo H, Liu G, Ding Y (2018) Moon-based Earth observation: scientific concept and potential applications. Int J Digit Earth 11(6):546–557
Guo H, Chen F, Qiu Y et al (2013) Fifty years of global space-borne Earth observation and China's development. Bull Chin Acad Sci Supplementary Issue:7–16
Guo H, Fu W, Li X et al (2014b) Research on global change scientific satellites. Sci China Earth Sci 57(2):204–215
Guo J, Han Y, Hwang C (2008) Analysis on motion of Earth's center of mass observed with CHAMP mission. Sci China Ser G: Phys Mech Astron 51(10):1597–1606
Huang Y, Wang MH, Mao JT (2004) Retrieval of upper tropospheric relative humidity by the GMS-5 water vapor channel: A study of the technique. Adv Atmos Sci 21:53–60
Jiang J (2013) The Booming Space Microwave Remote Sensing Technology of China. Bull Chin Acad Sci 28:33–39
Kim M-J, Ou M-L, Sohn E-H et al (2011) Characteristics of sea surface temperature retrieved from MTSAT-1R and in-situ data. Asia-Pac J Atmos Sci 47(5):421–427
Liu F (2009) GRACE satellite time-varying gravity field research. Dissertation, Chang'an University
Lu X (2005) The method of using satellite-to-satellite tracking data to recover gravity field and numerical simulation. Dissertation, Huazhong University of Science and Technology
Melzer BA, Subrahmanyam B (2017) Evaluation of GRACE mascon gravity solution in relation to interannual oceanic water mass variations. IEEE Trans Geosci Remote Sens 55(2):907–914
Metzler B, Pail R (2005) GOCE data processing: the spherical cap regularization approach. Stud Geophys Geod 49(4):441–462
National Satellite Meteorological Center, China Meteorological Administration (2013a) FY-3 satellite. http://nsmc.cma.gov.cn/. Accessed 10 Dec 2018
National Satellite Meteorological Center, China Meteorological Administration (2013b) Satellite operation. http://nsmc.cma.gov.cn/. Accessed 10 Dec 2018
National Satellite Ocean Application Service (2007) China maritime satellite application report (HY1B). 2007-1. http://www.nsoas.gov.cn/. Accessed 10 Dec 2018
National Satellite Ocean Application Service (2011) China maritime satellite application report (HY-2A). 2012-7. http://www.nsoas.gov.cn/. Accessed 10 Dec 2018
Neeck SP, Jucks KW, Lindstrom EJ et al (2008) The decadal survey tier 2 missions. In: Meynart R, Neeck SP, Shimoda H, Habib S (eds) Proceedings volume 1706, sensors, systems, and next-generation satellites XII, 71060A. SPIE remote sensing, Cardiff, Wales, 9 October 2008
Satellite Environment Center, Ministry of Environmental Protection (2010a) Environmental satellites. http://www.secmep.cn/. Accessed 10 Dec 2018

Sathiyamoorthy V, Sikhakolli R, Gohil BS et al (2012) Intra-seasonal variability in Oceansat-2 scatterometer sea-surface winds over the Indian summer monsoon region. Meteorol Atmos Phys 117(3):145–152

Veefkind JP, Aben I, McMullan K et al (2012) TROPOMI on the ESA Sentinel-5 precursor: a GMES mission for global observations of the atmospheric composition for climate, air quality and ozone layer applications. Remote Sens Environ 120:70–83

Visser PNAM (2010) A glimpse at the GOCE satellite gravity gradient observations. Adv Space Res 47(3):393–401

Wingham DJ, Francis CR, Baker S et al (2006) CryoSat: a mission to determine the fluctuations in Earth's land and marine ice fields. Adv Space Res 37(4):841–871

Zhou G (2010) Architecture of future intelligent Earth observation satellites (FIEOS) in 2010 and beyond. Technical report (June 1st 2001-November 31st 2001). Submitted to National Aeronautics and Space Administration Institute of Advanced Concepts (NASA-NIAC)

Wenxue Fu received the Ph.D. degree in SAR remote sensing from Nanjing University, Nanjing, China, in 2007. He is currently an Associate Professor with the Institute of Remote Sensing and Digital Earth, Chinese Academy of Sciences, Beijing, China. His research interests include radar interferometry for ground deformation retrieval and polarimetric SAR interferometry for forest parameter inversion.

Pei Chen received his Ph.D. degree in aerospace engineering from Beihang University, Beijing, China, in 2007. He is currently an associate professor at the school of astronautics, Beihang University. His current research activities comprise spacecraft navigation, GNSS application, and orbit determination.

Fang Chen is a professor at the Institute of Remote Sensing and Digital Earth, Chinese Academy of Sciences (CAS). Dr. Chen conducts interdisciplinary work combining, remote sensing, ecology, and other fields of study to assess spatial patterns of disaster risk.

4

Satellite Navigation: System Structure, Signals and Service Performance

Chuang Shi and Na Wei

Abstract Global navigation satellite systems (GNSSs) have been widely used in navigation, positioning, and timing. China's BeiDou Navigation Satellite System (BDS) would reach full operational capability with 24 Medium Earth Orbit (MEO), 3 Geosynchronous Equatorial Orbit (GEO) and 3 Inclined Geosynchronous Satellite Orbit (IGSO) satellites by 2020 and would be an important technology for the construction of Digital Earth. This chapter overviews the system structure, signals and service performance of BDS, Global Positioning System (GPS), Navigatsionnaya Sputnikovaya Sistema (GLONASS) and Galileo Navigation Satellite System (Galileo) system. Using a single GNSS, positions with an error of ~ 10 m can be obtained. To enhance the positioning accuracy, various differential techniques have been developed, and GNSS augmentation systems have been established. The typical augmentation systems, e.g., the Wide Area Augmentation System (WAAS), the European Geostationary Navigation Overlay Service (EGNOS), the global differential GPS (GDGPS) system, are introduced in detail. The applications of GNSS technology and augmentation systems for space-time geodetic datum, high-precision positioning and location-based services (LBS) are summarized, providing a reference for GNSS engineers and users.

Keywords BDS · GNSS augmentation systems · High-precision positioning · Real-time · LBS

C. Shi (✉)
School of Electronic and Information Engineering, Beihang University, Beijing, China
e-mail: shichuang@buaa.edu.cn

C. Shi · N. Wei
GNSS Research Center, Wuhan University, Wuhan, China

4.1 Introduction

The concept of Digital Earth was proposed by former US vice president Al Gore in 1998. At the 6th International Symposium on Digital Earth in Beijing, Digital Earth was defined as an integral part of other advanced technologies, including earth observation, geo-information systems, global positioning systems, communication networks, sensor webs, electromagnetic identifiers, virtual reality, and grid computation. Satellite navigation and positioning technology can provide precise position and time information, which are key elements of the Digital Earth.

In satellite navigation and positioning technology, the radio signals transmitted by navigation satellites are received by the user terminal. By measuring the time delay of the signal propagated from the navigation satellite to the receiver, navigation, positioning and timing services can be realized. Compared with conventional navigation and positioning techniques, satellite navigation and positioning technology can provide precise three-dimensional positions, velocity and time for users. It is an all-weather, all-time and globally available technology. Great progress has been made in recent decades, and many countries and consortia have established their own global navigation satellite systems. Global satellite navigation and positioning technology has been widely applied in navigation for vehicles, offshore ships, aero craft and aerospace vehicles and in the fields of geodesy, oil exploration, precision agriculture, precise time transfer, and earth and atmospheric sciences.

4.2 Global Navigation Satellite System

Before the advent of man-made satellites, navigation and positioning mainly depended on ground-based radio navigation systems that were developed during the Second World War such as LORAN and Decca Navigator, shown in Fig. 4.1. On October 4th 1957, the former Soviet Union (Union of Soviet Socialist Republics, or USSR) launched the first man-made satellite. Based on the Doppler shift of the radio signal, Dr. Guier and Dr. Wiffenbach from Johns Hopkins University successfully calculated the orbit of the satellite. This laid the foundation for the scientific idea of navigation and positioning with the use of man-made satellites. In 1958, the US military began to develop the first (generation) satellite navigation and positioning system in the world—the Transit navigation satellite system (TRANSIT), which was formally put into military use in 1964. The USSR also began to establish the CICADA system in 1965, and the first CICADA satellite was launched in 1967. Using the Doppler shift method, the first-generation satellite positioning system needed long-term observations to realize navigation and positioning, and the positioning accuracy was also unsatisfactory. To overcome these limitations, the joint development of a new generation satellite navigation system—the Global Positioning System (GPS)—by the US army, navy and air force was formally approved

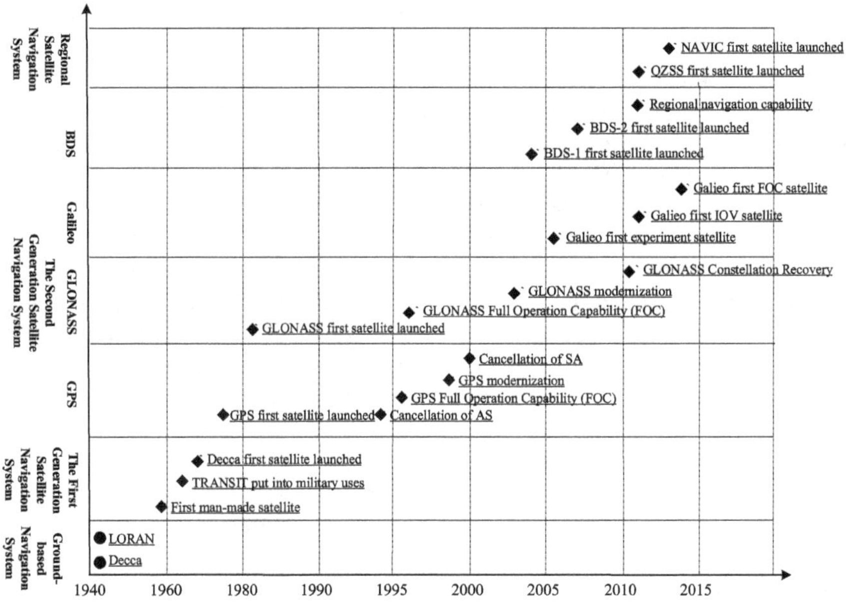

Fig. 4.1 Overview of the development of satellite navigation systems from the 20th century

by the United States Department of Defense (DoD), opening a new chapter for the development of satellite navigation systems.

The satellite navigation system was initially designed for military requirements. With the end of the Cold War, the growing demand for civil and commercial navigation became increasingly strong. Many countries in the world began to develop independent global navigation satellite systems (GNSSs), including the GPS developed by the US, the Globalnaya Navigatsionnaya Sputnikovaya Sistema (GLONASS) developed by Russia, the Galileo Navigation Satellite System (Galileo) established by the European Union (EU), and the BeiDou Navigation Satellite System (BDS) developed by China. Since the technical reserve and capital investment required for the GNSS development is rather large, some countries began to develop regional navigation satellite systems (RNSS) to meet the navigation and positioning demands in their own territory and the surrounding areas, for example, the Quasi-Zenith Satellite System (QZSS) of Japan and the Navigation with Indian Constellation (NAVIC) of India.

GNSSs have evolved from a single GPS constellation to multiple GNSS constellations. In the coming decades, the number of navigation satellites in orbit may increase to several hundreds. Therefore, the integration of multifrequency and multisystem GNSS data, compatibility of GNSS signals and interoperability between different GNSSs are the most important development directions.

Satellite navigation systems consist of three components: the space segment, the control segment (CS) and the user segment. The space segment comprises a constellation of navigation satellites that continuously broadcast ranging code and navigation message to users, and receive various information and commands from a ground monitoring system. The design of the navigation satellite constellation should ensure that four satellites are visible by any user at any time for positioning. The CS includes master control stations (MCSs), uplink stations and monitoring stations. The ground monitoring stations track navigation satellites and the MCSs collect observation data and calculate satellite orbit and clock errors, which are forecasted and formatted into navigation messages and uploaded to the navigation satellites through the uplink stations. The ground CS can also send various commands to the satellites through uplink stations for satellite orbit maneuver, atomic clock adjustment, fault recovery, or initiation of spare parts. The geometric distance between the navigation satellite and the receiver can be measured by the user with a GNSS receiver, and parameters such as the three-dimensional position, velocity and receiver clock errors can be obtained according to the satellite's location in space described by the ephemeris. As the main functions of these segments are similar for different satellite navigation systems, we ignore the common details in the following sections and introduce the unique features of each GNSS.

4.2.1 BDS

BDS, formerly known as COMPASS, is an independent global satellite navigation system developed and operated by China. As the third mature satellite navigation system after GPS and GLONASS, BDS provides high-quality positioning, velocity measurement, timing and short message services for global users. BDS has evolved from active positioning to passive positioning. A global passive positioning system will be established by 2020 (http://www.beidou.gov.cn).

Development of the BeiDou Navigation Satellite Demonstration System (BDS-1) was initiated in 1994. Two geosynchronous equatorial orbit (GEO) satellites were launched in 2000, and the regional double-satellite positioning system was established and put into operation. Based on the active-positioning scheme, positioning, timing, wide-area differential and short message communication services were provided for users in China. With the third GEO satellite launch in 2003, the system performance was further enhanced.

Development of the regional BeiDou Navigation Satellite System (BDS-2) began in 2004. As a passive-positioning system, BDS-2 can provide positioning, timing, wide-area differential and short message communication services for users in the Asia-Pacific region. By the end of 2012, the deployment goal of a regional satellite navigation system was accomplished, with a constellation of 5 GEO satellites, 5 inclined geosynchronous satellite orbit (IGSO) satellites and 4 medium earth orbit (MEO) satellites. On December 27th, 2012, it was officially declared that the BDS could provide regional positioning and navigation services with positioning accuracy of 10 m. China became the third country in the world with an independent satellite navigation system.

In a third step, the global BeiDou Navigation Satellite System (BDS-3) should be completed in 2020, with a constellation of 30 satellites (3GEO + 3IGSO + 24MEO). Both the GEO and IGSO satellites operate in orbits at an altitude of 35,786 km (BDS-ICD 2013). The inclination of the IGSO orbital plane is 55°. The altitude of the MEO satellites is 21,528 km, and the inclination is 55°, with a satellite orbit period of 12 h and 53 min.

BDS-3 has entered into a new era of global deployment with the introduction of new functions such as intersatellite links, and global search and rescue. The technical scheme of BDS-3 is fully forward compatible with that of BDS-2 and realizes performance improvement and service extension. By the end of 2018, the BDS-3 'basic system' comprising of 18 MEO and one GEO satellites was completed to provide services for users in China and neighboring countries along the Belt and Road. In-orbit validation has shown that the positioning accuracy is 10 m globally and 5 m in the Asia-Pacific area. By 2020, BDS-3 will be fully completed to provide global services and an integrated positioning navigation and timing (PNT) system should be set up by 2035.

The code division multiple access (CDMA) signal system is used by the BDS and the carrier signal is broadcasted at B1, B2 and B3 frequencies in L band. B1I and B3I were maintained and inherited from BDS-2, and a new open signal B1C was added and the B2 signal was also upgraded into the newly designed B2a signal, which replaces the original B2I signal and greatly improves the signal performance of BDS-3. The compatibility and interoperability with other GNSSs were also taken into account. A domestically developed high-precision rubidium and hydrogen atomic clock with better stability and smaller drift rate was equipped on the BDS-3 satellites, leading to significant improvement in the performance of the onboard time and frequency standards.

The intersatellite links in the Ka frequency band are equipped for the BDS-3 constellation, and two-way intersatellite precise ranging and communication is realized through use of phased-array antenna and other intersatellite link equipment. Mutual ranging and timing through intersatellite links allow for obtaining more measurements from multiple satellites to improve the observation geometry for autonomous orbit determination. The intersatellite measurement information can also be used to calculate and correct satellite orbit and clock errors for satellite-satellite-ground integrated precise orbit determination, improving the accuracy of satellite orbit determination and time synchronization. Both open and authorized services are provided by BDS-3. The open service provides free services for global users with a positioning accuracy of 10 m, velocity measurement accuracy of 0.2 m/s and timing accuracy of 10 ns. The authorized service provides authorized users with high-precision and reliable measurement of position, velocity and time, communication services, and system integrity information.

The basic BDS observations include pseudorange and carrier phase measurements. The pseudorange measurement is calculated by multiplying the speed of light with the transmission time of the GNSS ranging code from the satellite to the receiver, which comes from the correlation operation of the ranging code generated by the receiver clock with that generated by the satellite clock. The pseudorange reflects the distance between the satellite antenna phase center at the time when the GNSS signal is transmitted by the satellite and the receiver antenna phase center when the signal arrives. Its accuracy therefore depends on the code correlation accuracy. Currently, the noise of the pseudorange measurement is approximately 1%–1‰ of the code width.

The carrier phase measurement refers to the measurement of the navigation signal received from the satellite relative to the carrier phase generated by the receiver (the beat frequency phase) at the time of reception. Once the receiver is powered on, the fractional part of the beat frequency phase is measured and the changes in the integer number of carriers are counted. However, the initial integer number of carriers between the receiver and the satellite cannot be measured. Taking a complete carrier as one cycle, the unknown number of integer cycles is called the ambiguity. The initial measurements of the carrier phase include the correct fractional part and an arbitrary integer number of cycles at the starting epoch. At present, the accuracy of the carrier phase measurement recorded by electronic devices is better than 1% of the wavelength; that is, the carrier phase measurement accuracy is millimeter level.

Compared with the other existing GNSSs, the BDS has the following features: first, the space segment of the BDS is a hybrid constellation comprised of satellites in three kinds of orbits, and the anti-jamming and anti-spoofing capability is better due to more satellites in higher orbits, especially for the low latitude regions; second, the BDS is the first GNSS with signals broadcasted at three frequencies in the full constellation, which could improve service accuracy with a multifrequency combination signals; third, navigation and communication are innovatively integrated in the BDS, so that it can implement five major functions including providing real-time navigation and positioning, precise timing and short message communication services. The service performance of BDS are summarized in Table 4.1.

Table 4.1 Overview of BDS service performance

	BDS-1	BDS-2 (regional)	BDS-3 (global)
Service coverage	China and neighboring areas	Longitude: 84°–160° E, Latitude: 55° S–55° N	Global
Positioning accuracy	<20 m	Horizontal 25 m, Elevation 30 m	<10 m (three-dimensional)
Velocity accuracy	/	0.2 m/s	<0.2 m/s
Timing accuracy	100 ns for one-way, 20 ns for two-way	50 ns	20 ns

4.2.2 GPS

The GPS space segment consists of a constellation of 24 satellites, 21 operational satellites and 3 spare satellites, shown in Fig. 4.2. Four satellites are equally spaced in each of the six orbital planes with an orbit inclination of 55°. The difference between the ascending nodes of each orbital plane is 60° and the difference in the argument of latitude for satellites in the same orbital planes is 30°. This ensures that at least four GPS satellites can be visible at any time and any location around the world. The average orbital altitude of the GPS satellites is approximately 20,200 km, and their orbital period is approximately 11 h 58 min 2 s. For more information about GPS, please refer to http://www.gps.gov/.

The first GPS satellite was launched in 1978, and the constellation of 24 MEO satellites was completed in 1994. Based on the launch time, the GPS satellites can be divided into six different types, namely, BLOCK I, BLOCK II/IIA, BLOCK IIR, BLOCK IIR-M, BLOCK IIF and GPS III satellites. The CDMA modulation is also adopted for GPS satellites to broadcast carrier signal in the L1 band (1575.42 MHz) and L2 band (1227.60 MHz). The open civil C/A code is modulated on carrier L1, and the encrypted P(Y) code for military uses is modulated on carrier L2 (ICD-GPS-200J 2018).

To further expand the GPS civil market and better serve military demands, the GPS modernization program was promoted by the US government. As shown in Table 4.2, the modernization of the GPS navigation signal and satellite constellation includes:

(1) Broadcasting a new civil navigation signal and new military code (M code). The second civil pseudorange code L2C was introduced on BLOCK IIR-M satellites in 2005, and the third carrier frequency L5 (1176.45 MHz) was added on

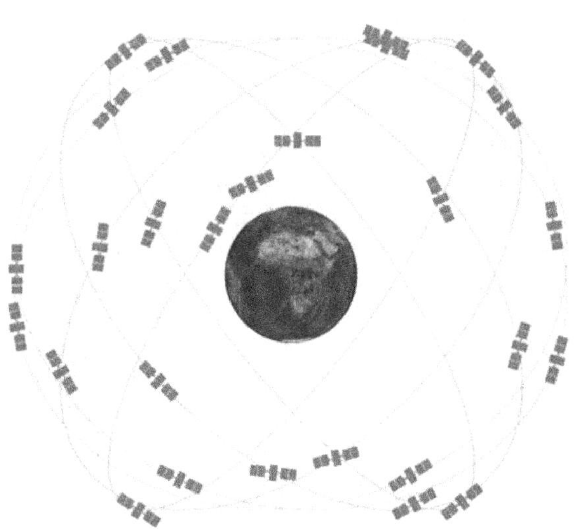

Fig. 4.2 Constellation of the GPS system.

Table 4.2 GPS satellite constellation and navigation signal modernization

	BLOCK I	BLOCK II/IIA	BLOCK IIR	BLOCK IIR-M	BLOCK IIF	GPS III
Civil code	C/A	C/A	C/A	L2C added	L5 added	L1C added
Military code	P(Y)	P(Y)	P(Y)	M code added		
Designed lifetime	4.5 years	7.5 years	7.5 years	7.5 years	12 years	15 years
Launch time	1978–1985	1990–1997	1997–2004	2005–2009	2010–2016	2016–present
	No SA ability				L5 added	No SA ability
						With laser prism reflector

BLOCK IIF satellites in 2010. In 2016, the GPS III satellites began to broadcast three GPS carrier frequencies (L1, L2 and L5) with four civil navigation codes (C/A, L2C, L5 and L1C), among which the L2C was mainly designed for commercial applications. L5 was developed to meet the demands of navigation users in the field of safety-of-life-related transportation and other high-precision applications, and L1C was designed for compatibility and interoperability between GPS and other GNSSs.

(2) Launching the new generation GPS III satellites to gradually replace the earlier satellites. The GPS III satellites were no longer able to implement the Selective Availability (SA) policy and a laser prism reflector was carried onboard to separate the satellite orbit and clock errors. The lifespan of the GPS satellites was also extended.

Until 2016, the GPS ground control segment consisted of one MCS, one backup MCS, 15 globally distributed monitoring stations, 11 uplink stations and the auxiliary communication network, shown in Fig. 4.3. Its MCS was located in Colorado, US. The ground control segment upgrade was included in the GPS modernization program and consisted of the following main aspects: (1) the Legacy Accuracy Improvement Initiative (L-AII) plan completed in 2008; ten GPS monitoring stations that belonged to the National Geospatial-Intelligence Agency (NGA) were added to the ground monitoring network to improve the forecasting accuracy of the GPS broadcast ephemeris; (2) the Architecture Evolution Plan (AEP) for MCS IT upgrade and the Launch and early orbit, Anomaly resolution, and Disposal Operations (LADO) plan for monitoring out-of-operation satellites in 2007; and (3) the Next Generation Operational Control System (OCX) plan implemented in 2010.

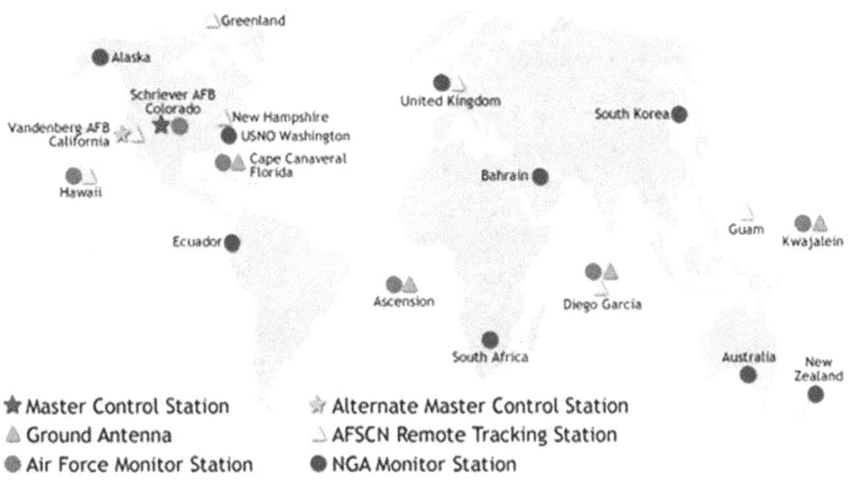

Fig. 4.3 GPS ground control segment.

The major function of the GPS user segment, including GPS receivers and handheld terminals, is to track GPS satellites and compute the three-dimensional positioning, velocity and time for users. Users can receive two types of GPS positioning services: standard positioning service (SPS) and precise positioning service (PPS), shown in Table 4.3. SPS is free for all users. The positioning and timing services are obtained with C/A code on L1 and the broadcast ephemeris. PPS is aimed at serving the military and authorized users, and the positioning and timing services are obtained using the ranging code modulated on both L1 and L2 (Grimes 2007).

4.2.3 GLONASS

GLONASS was developed by the USSR in 1976 and is now operated by Russia. With the first GLONASS satellite launched on October 12, 1982, the full constellation was completed and was put into operation at the beginning of 1996. However, due to the short satellite lifespan of only 2–3 years on average and the lack of adequate funding to launch supplementary satellites after the economic recession, there were only six operational GLONASS satellites on orbit in 2011, which severely impacts the normal use of GLONASS. With the recovery of the Russian economy, the GLONASS modernization program was initiated. At the end of 2011, the full 24-satellite constellation was restored for independent navigation and positioning capability.

The space segment of GLONASS consists of 21 operational satellites and 3 backup satellites, which are evenly distributed over three orbital planes with an inclination of 64.8°. The longitude of the ascending node of each plane differs by 120° from

Table 4.3 Performance of GPS SPS and PPS

	Service performance		SPS	PPS	
	Coverage		Global (civil)	Global (authorized)	
Signal in space	Accuracy			Single frequency	Double frequency
	Ranging accuracy			5.9 m	6.3 m
	Accuracy of ranging rate			0.006 m/s	0.006 m/s
	Accuracy of ranging acceleration			0.002 m/s^2	0.002 m/s^2
	Accuracy of timing			40 ns	40 ns
	Integrity (95%) SIS URE	Alarm threshold	150 m	150 m	
		Warning threshold	8 s	8 s	
		Integrity risk	1×10^{-5}/h	1×10^{-5}/h	
	Continuity risk		SIS: 0.0002/h	SIS: 0.0002/h	
	Single slot availability		0.957	0.957	
Performance indicators	Accuracy (95%)		Horizontal: 9 m Vertical: 15 m		
	PDOP availability		Global average: PDOP ≤ 6 (98%) Worst case: PDOP ≤ 6 (88%)	Global average: PDOP ≤ 6 (98%) Worst case: PDOP ≤ 6 (88%)	
	Accuracy availability		Horizontal: 17 m Vertical: 37 m		

plane to plane. In each orbit plane, there are 8 satellites separated by 45° in argument of latitude (ICD-GLONASS 2008). At an altitude of 19,100 km, the orbital period is 11 h 15 min and 44 s. In September 2016, the number of operational satellites in orbit was increased to 27, 24 of which are GLONASS-M and GLONASS-K1 satellites with full operational capability. For more information about GLONASS, please refer to https://www.glonass-iac.ru/.

Frequency division multiple access (FDMA) modulation is used by GLONASS; thus, different satellites are distinguished by different frequencies. The frequencies of the civil signals G1 and G2 broadcasted by GLONASS satellites are as follows:

$$\begin{cases} G1 : f_{K1} = 1602\,\text{MHz} + 0.5626\,\text{MHz} \times K \\ G2 : f_{K2} = 1246\,\text{MHz} + 0.4375\,\text{MHz} \times K \end{cases} \tag{4.1}$$

where $K = -7, \ldots 0.6$ is the frequency number of each satellite. For any satellite, $f_{K1}/f_{K2} = 9/7$. The frequencies of carriers G1 and G2 for military use are different than those for civil use. Pseudo-random-noise code is modulated on the carrier signal and is the same for each set of frequencies. Similar to GPS, the civil code C/A was initially modulated only on carrier G1 whereas the military code P was modulated on both carriers G1 and G2. Later, the C/A code was also modulated on carrier G2 of GLONASS-M satellites. In contrast to GPS, the GLONASS P code is not encrypted.

The original intention to adopt the FDMA system in GLONASS was to enhance the anti-jamming capability. However, this strategy prevented the promotion of commercialization of GLONASS. To improve the compatibility and interoperability with other GNSSs, GLONASS began to broadcast the CDMA signal. For example, the first GLONASS-K1 satellite launched in 2011 broadcasted FDMA signals in the G1 and G2 bands and the new CDMA signal in the G3 band (1202.025 MHz), marking the first step of GLONASS signal modernization. In the future, the development of GLONASS CDMA signals will mainly focus on the G1 and G2 bands. As an improved version of GLONASS-K1, the GLONASS-K2 satellite could broadcast civil FDMA ranging code in G1 and G2 as well as the civil CDMA ranging code in G1, G2 and G3. GLONASS constellation modernization also includes stability and performance improvement of the onboard atomic clock, satellite lifespan extension, and introduction of intersatellite laser ranging (shown in Table 4.4).

The GLONASS system control center (SCC) is located in Krasnoznamensk, Moscow. The GLONASS time reference is maintained by the central clock (CC-M) in Schelkovo. Five telemetry, tracking and command (TT&C) centers are evenly

Table 4.4 GLONASS constellation modernization.

		GLONASS	GLONASS-M	GLONASS-K1	GLONASS-K2
Civil signal		G1OF	G1OF/G2OF	G1OF/G2OF G3OC	G1OF/G2OF G1OC/G2OC G3OC
Military signal		G1SF/G2SF	G1SF/G2SF	G1SF/G2SF	G1SF/G2SF G1SC/G2SC
Designed lifetime		3.5 years	7 years	10 years	10 years
Launch time		1982–2005	2003–2016	2011–2018	2017~
Clock stability		5×10^{-13} $\sim 1 \times 10^{-13}$	1×10^{-13} $\sim 5 \times 10^{-14}$	1×10^{-13} $\sim 5 \times 10^{-14}$	1×10^{-14} $\sim 5 \times 10^{-15}$
Modulation mode		FDMA	FDMA	FDMA/CDMA	FDMA/CDMA
Intersatellite links:	RF	–	+	+	+
	Laser	–	–	–	+

Note 'O' is the open signal, 'S' is the precision signal, 'F' is FDMA and 'C' is CDMA

Table 4.5 Overview of GLONASS service performance

Performance indicator			Performance specification
Signal-in-space	Ranging error for any satellite		18 m
	Ranging velocity error for any satellite		0.02 m/s
	Ranging acceleration error for any satellite		0.007 m/s^2
	RMS ranging error for all satellites		6 m
Service	Coverage		From the earth's surface to an altitude of 2000 km
	Positioning accuracy (95%)	Horizontal	5 m (global average) 12 m (global average)
		Vertical	9 m (global average) 25 m (global average)
	Timing accuracy		$\leq$700 ns
	Availability (95%)	Horizontal	12 m (global average $\geq$ 99%, worst case $\geq$ 90%)
		Vertical	25 m (global average $\geq$ 99%, worst case $\geq$ 90%)
	Reliability	Fault rate	$\leq$3 times/year
		Reliability	99.97%

distributed in Saint Petersburg, Schelkovo, Yenisseisk, Komsomolsk and Ussuriysk in Russia. Although the GLONASS TT&C stations are not distributed worldwide, a high-accuracy broadcast ephemeris can be generated by the ground control segment because the longitudinal span of the Russian territory is large. In addition, some TT&C stations are also equipped with a laser station (LS) and other Monitoring and measuring stations (MS). The service performance of GLONASS system are summarized in Table 4.5.

4.2.4 Galileo

Galileo was developed in a collaboration between the European Union and the European Space Agency (ESA). Galileo is the first global navigation satellite system designed for civil uses. The space segment of Galileo consists of 24 operational satellites and 6 spare satellites, which will be positioned on three orbital planes with an inclination of 56°, and the ascending nodes on the equator are separated by 120°. The orbital altitude is 23,222 km, and the orbital period is approximately 14 h (ICD-Galileo 2008).

The development of Galileo can be divided into three phases: the development system testbed, in-orbit validation (IOV) and full operational capability (FOC). During the development system testbed phase, two experimental satellites, GIOVE-A and

GIOVE-B, were launched in 2005 and 2008, respectively (known as the Galileo In-Orbit Validation Element, GIOVE). Later, four Galileo-IOV satellites were launched in 2011 and 2012. In 2014, the Galileo-FOC satellites began to be launched.

The ground control segment of Galileo comprises two parts: the ground mission segment (GMS) and the ground control segment (GCS). The GMS is mainly responsible for processing observations to generate broadcast ephemeris. One ground control center (GCC) is located in Fucino, Italy, and is mainly responsible for monitoring the satellite constellation along with the GCC in Oberpfaffenhofen, Germany. These two GCCs are responsible for coordination and control of the TT&C stations, several uplink stations (ULS) and the Galileo sensor stations (GSSs) distributed worldwide, to maintain routine operation of the control segment. Galileo attaches great importance to system augmentation and integrity, which helps ensure the positioning accuracy and reliability in the fields of aviation and other safety-of-life applications.

Galileo makes use of the CDMA system to broadcast carrier signals on four frequencies: E1, E5a, E5b and E6. Five types of services are provided by Galileo for users: the free open service (OS) similar to GPS SPS, the safety-of-life service (SoLS), the commercial service (CS), the public regulated service (PRS) and the search and rescue (SAR) service. Signal E1 supports OS/CS/SoL/PRS services, E6 supports CS/PRS services, E5a supports OS services, and E5b supports OS/CS/SoL services. The service performance of Galileo system are summarized in Table 4.6.

With the modernization of GPS and GLONASS and the deployment of BDS and Galileo, the GNSS constellations has developed from approximately 30 GPS

Table 4.6 Overview of Galileo service performance

Satellite self-standing service			OS	CS	SoLS	PRS
Service	Coverage		Global	Global	Global	Global
	Positioning (95%)	Single frequency	Horizontal: 15 m Vertical: 35 m		–	Horizontal: 15 m Vertical: 35 m
		Double frequency	Horizontal: 4 m Vertical: 8 m		Horizontal: 4 m Vertical: 8 m	Horizontal: 6.5 m Vertical: 12 m
		Timing	30 ns		30 ns	30 ns
	Integrity	Alarm threshold	–		Horizontal: 12 m Vertical: 20 m	Horizontal: 20 m Vertical: 35 m
		Alarm time			6 s	10 s
		Integrity risk			$3.5 \times 10^{-7}/150$ s	$3.5 \times 10^{-7}/150$ s
	Continuity		–	–	$1 \times 10^{-5}/15$ s	$1 \times 10^{-5}/15$ s
	Availability	Available accuracy	99.8%	99.5%	99.8%	99.5%
		Available integrity	–	–	99.5%	99.5%

Table 4.7 Summary of BDS/GPS/GLONASS/Galileo system

	GPS	GLONASS	Galileo	BDS
First Launch	1978-02-22	1982-10-12	2005-12-28	2017-11-05
FOC	1995-07-17	1996-01-18	/	/
Service type	Military/civil	Military/civil	Commercial/open	Military/civil
No. of designed satellites	24	24	30	30
No. of orbital planes	6	3	3	3 (MEO)
Orbital inclination	55°	64.8°	56°	55° (MEO)
Orbital altitude	20,200	19,100	23,222	21,528 (MEO)
Orbital period	11 h 58 m	11 h 15 m	14 h 04 m	12 h 53 m (MEO)
Coordinate system	WGS84	PZ-90	GTRF	BDCS
Time system	GPST	UTC(SU)	GST	BDT
Modulation mode	CDMA	FDMA	CDMA	CDMA
Frequencies	L1:1575.42 L2:1227.60 L5:1176.45	G1:1602.00 G2:1246.00 G3:TBD	E1:1575.42 E5a:1176.45 E5b:1207.14 E6:1278.75	B1:1575.42 B2:1176.45 B3:1268.52

satellites in the early stage to more than 100 GNSS satellites in September 2016, summarized in Table 4.7. RNSSs such as QZSS and NAVIC are also under development. As shown in Fig. 4.4, the global coverage and availability of satellite navigation system signals have been improved.

In addition, the frequencies and types of GNSS satellite signals are becoming increasingly abundant (as shown in Table 4.7). For example, the second civil ranging code L2C and the third frequency L5 have been gradually provided by modernized GPS satellites. In the future, GLONASS will simultaneously broadcast FDMA, CDMA, as well as the third frequency signal G3. BDS provides signals in three frequencies of the full constellation and Galileo could broadcast carrier signals on four frequencies and 10 ranging codes. There are over 75 types of measurements (Gurtner and Estey 2013). To meet the ever-growing demand for GNSS civil applications, the third frequency signals L5 and G3 that are used for safety-of-life applications were designed by GPS and GLONASS, respectively; as a civil GNSS, Galileo gave high priority to aviation, safety-of-life and SAR applications at the beginning of its development.

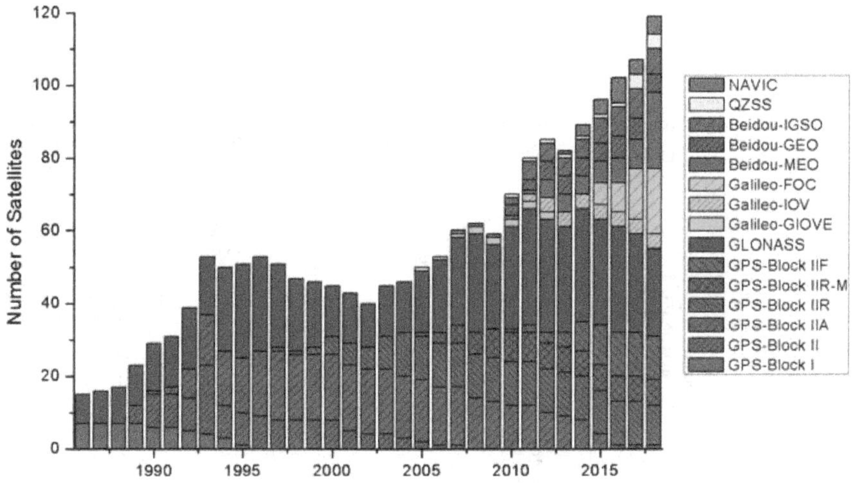

Fig. 4.4 The number of in-orbit satellites in GNSSs and RNSSs

4.3 GNSS Augmentation Systems

As described in Sect. 4.2, the accuracy of GNSS is rather limited and cannot meet the required positioning accuracy, time latency, reliability and integrity needs of higher-level users. The GNSS differential positioning technique and GNSS augmentation system strategy address this issue. In GNSS differential positioning, the observations of GNSS reference stations are used to model the error sources such as the ionospheric, tropospheric, satellite orbit and clock errors. These errors are then mitigated from the observation of users in real-time or post-processed mode to improve the accuracy and reliability of positioning. To meet the demands of different users, several different kinds of GNSS high-accuracy and real-time positioning systems have gradually been developed, including the wide-area differential augmentation system, the global/wide-area precise positioning system, the local area differential augmentation system and the local area precise positioning system (summarized in Table 4.8).

4.3.1 Wide-Area Differential Augmentation System

In the wide-area differential augmentation system, GNSS satellites are monitored with a ground tracking network and the raw observations are transferred to the master control center through communication links. The master control center models the errors of the GNSS raw observations and divides the errors into satellite orbit, clock, and ionospheric errors, which are formatted and broadcasted to users in the service region through communication links. The positioning accuracy can be improved by

Table 4.8 Overview of GNSS augmentation systems

	Reference station	Processing principle	Broadcast link	Broadcast format	Performance	Typical systems
Wide-area differential technique	Ranges of several thousand kilometers with several dozen reference stations	Differencing in the state domain, with corrections generated for orbit and clock errors, grid ionospheric VTEC values based on pseudorange measurements (assisted by the carrier phase)	GEO satellites or radio beacons	RTCA	Single-frequency pseudorange differential positioning, Precision: 3 m without initialization	WAAS EGNOS MSAS GAGAN SDCM
Global precise positioning technique	~100 globally reference stations	Differencing in the state domain, with corrections generated for orbit and clock errors based on carrier phase measurements	GEO satellites or through the Internet	SOC or User-defined protocol	Double-frequency carrier phase differential positioning, Precision: decimeter-level with initialization of approximately 20 min	GDGPS StarFire OmniStar SeaSTAR Veripos RTX
Local-area differential technique	Range of a few tens of kilometers with several reference stations	Differencing in the observation domain, with pseudorange corrections generated based on pseudorange measurements (assisted by the carrier phase)	U/VHF data link	RTCA or RTCM	Single-frequency pseudorange differential positioning Precision: submeter level without initialization	LAAS
Local-area precise positioning technique	Range of a few tens of kilometers with several reference stations and can be extended to larger regions with several thousand reference stations	Differencing in the observation domain, with regional error correction parameters generated in VRS, FKP, MAC and other processing modes based on carrier phase measurements. Users in the region can apply for such correction information	GPRS/CDMA communication link broadcasting	RTCM	Double-frequency carrier phase differential positioning Precision: centimeter level with an initialization of approximately 1–2 min	CORS

using the wide-area differential corrections. With uniform precision over broad coverage, the positioning accuracy of the wide-area differential augmentation system is independent of the distance between the user and the reference station. Several wide-area differential augmentation systems have been established worldwide, e.g., the Wide Area Augmentation System (WAAS) of the Federal Aviation Administration (FAA), the European Geostationary Navigation Overlay Service (EGNOS), the Multi-functional Satellite Augmentation System (MSAS) of Japan, the GPS-Aided Geo Augmented Navigation (GAGAN) of India and the System Differential and Correction Monitoring (SDCM) of Russia.

4.3.1.1 WAAS

The GPS SPS could not meet the higher accuracy, integrity, continuity and availability demands of users in aviation and other fields. As a result, the FAA initiated the WAAS program. As a satellite-based augmentation system (SBAS) to serve North America, WAAS is aimed at providing GPS differential correction information through GEO satellites to augment the GPS SPS. In addition to applications in the field of aviation, the WAAS has also been widely applied to support PNT services.

As illustrated in Fig. 4.5, the WAAS currently consists of 3 wide-area master stations (WMS), 38 wide-area reference stations (WRS), 4 Ground Uplink Stations and

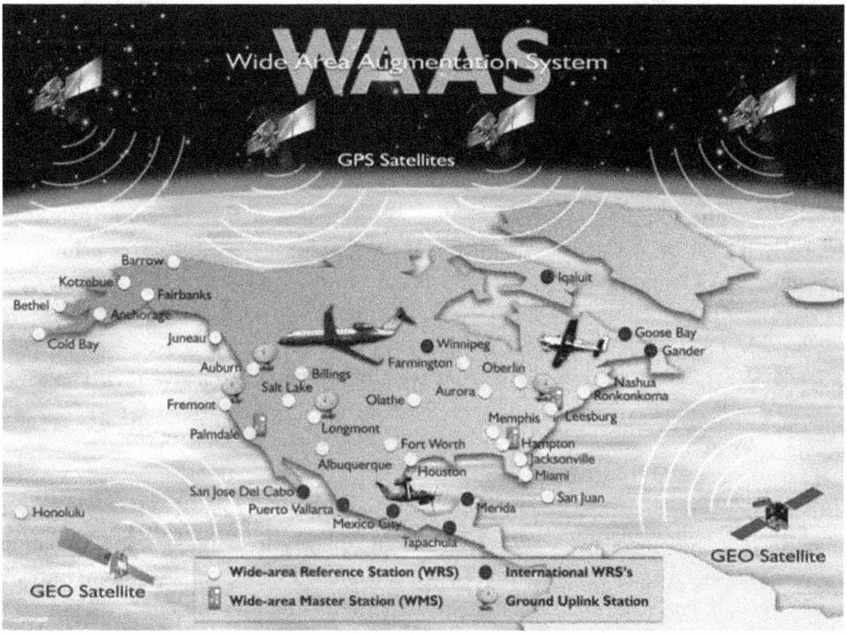

Fig. 4.5 Diagram of WAAS.

GEO satellites. The WMS is responsible for calculating the differential corrections and monitoring the system integrity. The WAAS data processing center receives real-time GPS observations from the WRS and computes differential correction vectors using the RTG (Real Time GIPSY) software package developed by the JPL based on GIPSY-OASIS. The corrections include the satellite orbit error, satellite clock error and the ionospheric error. Since these differential corrections are expressed as vectors, the precision of positioning within the areas covered by the wide-area differential system is equivalent, in contrast to the local-area differential system. The corrections are uploaded to GEO satellites through uplink stations and broadcast to users in RTCA format to improve positioning accuracy. The nominal positioning accuracy of WAAS (with 95% reliability) is better than 7.6 m in the horizontal and vertical directions. The horizontal and vertical positioning accuracy of WAAS are better than 1.0 m and 1.5 m, respectively, in most regions adjacent to the US, Canada and Alaska.

4.3.1.2 EGNOS

EGNOS is a joint program of the European Space Agency (ESA), the EU and the European Aviation Safety Agency (EASA). The working principle of EGNOS is similar to that of WAAS. The difference is that EGNOS broadcasts differential corrections and integrity information for GPS as well as the differential corrections for GLONASS. The EGNOS differential correction information is calculated using software developed from BAHN, the ESA-owned precise positioning and orbit determination software, and broadcasted by GEO satellites through the L band. The EU is considering extending the broadcast coverage from the EU to other regions, including countries neighboring the EU and Africa.

The EGNOS ground network consists of 39 ranging integrity monitoring stations (RIMSs), 4 mission control centers (MCCs) and 6 navigation land earth stations (NLESs). The 4 MCCs are located in Torrejon, Spain, Gatwick, England, Langen, Germany, and Ciampino, Italy. EGNOS presently provides three types of service: (1) free open service since October 2009, with positioning accuracy of 1–2 m; (2) safety-of-life service since March 2011, with positioning accuracy of 1 m; and (3) data access service since July 2012, with positioning accuracy better than 1 m.

4.3.1.3 MSAS

MSAS was jointly developed by the Japan Civil Aviation Bureau (JCAB) and Japan's Ministry of Land, Infrastructure and Transport, mainly to provide communication and navigation services for Japanese aviation users. MSAS covers all flight service areas of Japan, and can broadcast meteorological information to mobile users in the Asia-Pacific region. The space segment of MSAS consists of two multifunctional transport satellites (MTSats), which are second generation Himawari satellites, the geostationary meteorological and environmental survey satellite developed by Japan.

The two MTSats are positioned at 140° E and 145° E. The Ku and L bands are the two frequencies; the Ku band is mainly used to broadcast high-speed communication information and meteorological data. Similar to the GPS L1 frequency, the L band is mainly used for navigation services. The working principle of MSAS is similar as those of WAAS and EGNOS, and the RTG software authorized by JPL is used for data processing. From its initial operation in September 2007, MSAS remarkably improved the navigation service performance for Japanese airports located on remote islands and met the precision demand for the nonprecision approach specified by the International Civil Aviation Organization (ICAO).

4.3.1.4 GAGAN

The GAGAN system was jointly developed by the Indian Space Research Organization (ISRO) and the Airports Authority of India (AAI). The space segment of GAGAN consists of two GEO satellites positioned over the Indian Ocean. Two bands are applied in GAGAN: the C band is used for TT&C application and the L band is used to broadcast navigation information. The frequency of the L band is identical to that of the GPS L1 (1575.42 MHz) and L5 (1176.45 MHz); thus, GAGAN is compatible and interoperable with GPS. The GAGAN signal covers the whole Indian continent, providing users with GPS signals and differential corrections to improve the GPS positioning accuracy and reliability for Indian airports and other aviation applications. The key technique and core algorithm of GAGAN is also based on technical support from the JPL. The ground segment of GAGAN consists of a master station located in Bangalore, an uplink station and eight reference stations located in Delhi, Bangalore, Ahmedabad, Calcutta, Jammu, Port Blair, Guwahati and Thiruvananthapuram.

4.3.1.5 SDCM

Serving as the GLONASS satellite navigation augmentation system, Russia began developing the SDCM system in 2002 with an aim of providing differential augmentation information for GLONASS and other GNSSs. The space segment of SDCM consists of three GEO satellites, also called the Russian civil data relay (Luch/Loutch) satellites. The three satellites are known as Luch-5A, Luch-5B and Luch-5, located at 167° E, 16° W and 16° W, respectively.

4.3.2 Global Differential Precise Positioning System

The global differential precise positioning technique was developed from the wide-area differential GNSS and the precise point positioning technique. In global differential positioning systems, the GNSS pseudorange and carrier phase observations are collected by globally distributed GNSS dual-frequency monitoring stations and transferred to data processing centers through a real-time data transmission network to calculate the real-time precise satellite orbit, clock error and ionospheric corrections. Using the corrections, a user could realize decimeter to centimeter level precise positioning around the world. The wide-area differential augmentation system mainly serves navigation users with the pseudorange observations whereas the global precise positioning system mainly targets positioning users with the carrier phase observations. Well-known established representative global differential precise positioning systems include the global differential GPS (GDGPS) system applied for satellite orbit determination, scientific research and high-end commercial services, the StarFire system developed by NavCom, OmniSTAR and SeaStar by Fugro, CenterPoint RTX by Trimble and Veripos by Subsea7.

4.3.2.1 GDGPS

GDGPS is the global precise positioning system developed by the National Aeronautics and Space Administration (NASA) of the United States. As shown in Fig. 4.6, the ground monitoring network of the GDGPS consists of more than 200 real-time monitoring stations all over the world. These monitoring stations transmit real-time data to the GDGPS processing center at 1 Hz frequency. Among the tracking stations, over 75 monitoring stations belong to the JPL. They are evenly distributed worldwide and equipped with three-frequency receivers. The time latency of the GDGPS from receiving observations to generating and broadcasting real-time differential correction products is approximately 5 s. The real-time differential correction products can be broadcasted in a variety of ways, including through the Internet, a VPN, T1, frame relay, modem and satellite broadcasting.

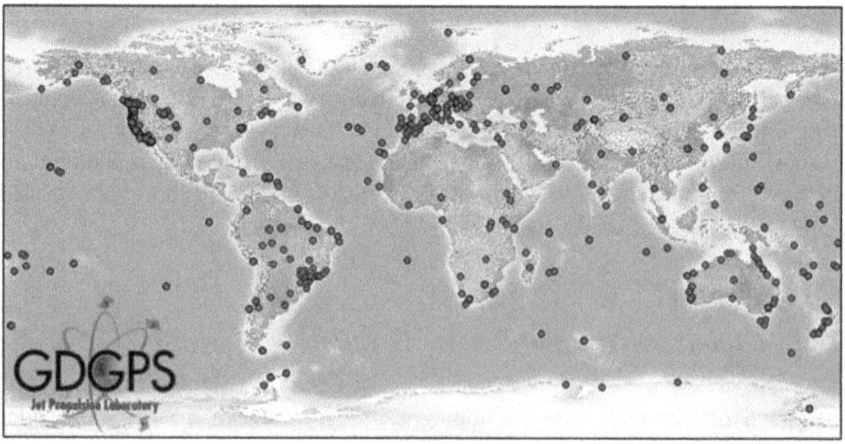

Fig. 4.6 The real-time tracking network of the GDGPS.

The JPL-developed RTG software is adopted in the GDGPS, with the state-square approach proposed by the JPL as its core algorithm. Based on the real-time dual-frequency GPS observation data collected by GDGPS monitoring stations worldwide, precise satellite orbit and clock errors are determined and compared with relevant parameters in the GPS broadcast ephemeris to generate differential corrections. The corrections are broadcasted to users for precise point positioning. The GDGPS can provide decimeter-level positioning and subnanosecond-level time transfer for dual-frequency GPS receivers over the globe. Compared with traditional differential positioning services, the positioning accuracy has been improved by one order of magnitude. The single-frequency users can also use the global ionospheric TEC maps provided by the GDGPS to implement ionospheric correction and improve positioning accuracy.

4.3.2.2 StarFire

In the early stage, StarFire was designed to provide independent wide-area differential augmentation services for precision agriculture in North America, South America, Europe and Australia. The early StarFire system was similar to WAAS and EGNOS, except that StarFire users must be equipped with high-quality dual-frequency GPS receivers to eliminate ionospheric delay, and should adopt the wide area correction transform (WCT) technique that was developed on the basis of NavCom.

In 2001, an agreement was reached between NavCom and NASA/JPL to upgrade StarFire into a global dual-frequency GPS precise positioning system based on the RTG technique. Continuous real-time positioning services with subdecimeter accuracy can be accessed anywhere and anytime around the world. In addition to the RTG technique, StarFire can access observation data from the NASA/JPL global monitoring network to augment the StarFire ground monitoring network. In addition, StarFire makes use of the International Maritime Satellite (INMARSAT) to broadcast differential signals to global users through the L band. StarFire users equipped with L band communication receivers can track and observe GPS satellites and receive the differential correction signals broadcasted by INMARSAT.

RTG/RTK is a new real-time differential positioning mode recently launched by NavCom. The disadvantage of a relatively long initialization time in RTG can be overcome by using RTK. If lock-lose or communication interruption of data links occur during RTK, real-time positioning services can be continuously provided by the RTG with centimeter-level positioning accuracy. After restoring the signal tracking and data link communication in RTK, the positioning result of the RTG can be used as the initial value for rapid searching and integer ambiguity resolution. The disadvantage is that at least two RTG/RTK combined dual-frequency receivers are required on the user side for real-time positioning.

4.3.2.3 OminiSTAR/SeaStar

OmniSTAR and SeaStar are real-time global differential systems developed by the Fugro company. OmniSTAR is mainly used in land and aviation applications and SeaStar was established to meet the demands for high kinematic positioning in marine applications. OmniSTAR currently provides four types of differential GPS positioning services with different accuracies, VBS, HP, XP and G2, among which the G2 service can support both GPS and GLONASS. OmniSTAR has been widely applied in agriculture, GIS, aviation, surveying and mapping, asset tracking and monitoring and, thus, occupies a considerable market share in differential GPS services.

SeaStar primarily serves the offshore oil and gas exploitation industry to meet the demands for submeter and decimeter-level kinematic positioning with high accuracy and reliability under special circumstances. It can provide G2, XP2, SGG and standard L1 services for GPS and GLONASS, as well as XP service for GPS. The latest G4 service simultaneously supports GPS, GLONASS, BeiDou and Galileo.

4.3.2.4 CenterPoint RTX

CenterPoint RTX (Real-Time eXtended) is a global real-time differential system developed by the US company Trimble. It provides worldwide precise positioning services with a horizontal accuracy better than 4 cm for GPS, GLONASS and QZSS. Using the corrections broadcasted by the L-band satellite, the initialization time needed for positioning is less than 5 min.

4.3.2.5 Veripos

Established by the Subsea7 company, Veripos consists of more than 80 reference stations all over the world. The two control centers of Veripos are located in Aberdeen, Britain and Singapore. The Veripos Apex, Veripos Apex^2, Veripos Ultra and Veripos Ultra^2 can provide services with a positioning accuracy better than 10 cm, and Veripos Standard and Veripos Standard^2 can provide a positioning accuracy better than 1 m. The latest Veripos Apex^5 can provide augmentation services for GPS/GLONASS/Galileo/BDS/QZSS with a horizontal positioning accuracy better than 5 cm (within a 95% confidence level).

4.3.3 Local Area Differential Augmentation System

The local area augmentation system (LAAS) at airports is a typical local area differential augmentation system. Based on GPS real-time differential corrections, LAAS was established as an all-weather precision approach and landing system. It consists of reference stations, a central processing station and airborne equipment. GPS satellites are continuously tracked by several reference stations located around the airport, and the central processing station receives the GPS observations to generate pseudorange differential corrections and integrity and precision approach and landing data, which are encoded and broadcasted to the airplane through VHF data links. Based on the GPS observations, differential corrections and integrity information broadcasted by LAAS, the airborne equipment can improve the navigation accuracy, integrity, continuity and availability to realize precision approach category I (CAT I) along a specified path. The ultimate goal of an LAAS is to provide CAT II and CAT III.

4.3.4 Local Area Precise Positioning System

In the local area precise positioning system, several GNSS reference stations are established in a certain region (district, city or country) and high-accuracy positioning service is provided to users within its coverage by taking advantage of differential corrections through a wired/wireless real-time data communication link. The local area precise positioning system can be categorized into two operational modes, single reference station mode and multiple reference stations mode. In single reference station mode, a reference station directly broadcasts high-precision carrier phase measurements to users at the rover station. The rover station receives the measurements and realizes precise positioning based on the differential positioning technique. The positioning accuracy of single reference station mode is at the centimeter level, which can meet the demands of applications within a small area.

The continuously operating reference station (CORS) system is a local area precise positioning system operated in the multiple reference station mode. CORS consists of continuously operating GNSS reference stations, which are interconnected by computer, data communication and the Internet. The observation data (carrier phase and pseudorange) at CORS reference stations are transmitted to the data processing center through the communication link in real-time. The data from the reference network are then uniformly processed to calculate and model the real-time corrections for various GNSS errors within the region, such as the satellite orbit/clock error an ionospheric and tropospheric error. The corresponding observation data and GNSS error model are broadcasted by the data processing center to users at the rover station for high-accuracy positioning. Many countries have established their own CORS systems at the national level, including the US, Germany, England, Australia, Japan, and Canada. Brief introductions to the US CORS and EPN in Europe are provided as representative examples.

4.3.4.1 US CORS

The establishment of CORS in the US was led by the National Geodetic Survey (NGS). More than 200 agencies and organizations have been involved in this program. The three largest CORS networks include the national CORS network, the operational CORS network and the California CORS network. In 2015, the US CORS consisted of more than 2000 reference stations. Most of the stations are distributed in American. However, several stations are located in Canada, Mexico, Central America and North America. The US CORS provides users with coordinates under the International Terrestrial Reference Frame (ITRF) and the 1983 North American Datum (NAD83), as well as raw observations and satellite orbit products. Real-time differential positioning service is also available in some areas, e.g., the San Diego real-time network.

4.3.4.2 EPN in Europe

The EUREF permanent network (EPN) in Europe is a cooperative regional continuously operating network established by the European Commission of the IAG. The EPN was composed of 250 permanent reference stations in 2016. The workflow of the EPN is as follows: the reference stations are divided into several subnetworks with independent system operation centers. Several system operation centers constitute a regional data center, and the data from regional data centers are gathered into the European regional center, which transfers the data products to the IGS data center, regional data centers and various kinds of users. The EPN provides users with centimeter-level coordinates and velocity in ITRF and EUREF, as well as zenith tropospheric products for the reference stations. In addition, the EPN can be applied to monitor crustal deformation, sea level changes, and in numerical weather prediction (NWP).

GNSS augmentation systems have achieved significant developments in the aspects of the accuracy, integrity, coverage and differential mode. The positioning accuracy of GNSS augmentation systems has improved from meter-level, as for WAAS, EGNOS and MSAS, to real-time decimeter-level and post-processed centimeter-level, e.g. StarFire and OmniStar. The GNSS augmentation system has been extended from regional coverage to seamless global coverage. The early GNSS augmentation systems provided users with correction based on one differential approach whereas the current system can provide users with high-precision positioning with corrections derived from multiple differential approaches. As an effective supplement to GNSS, the GNSS augmentation systems have greatly improved the GNSS SPS performance to meet the ever-growing demands for integrity, continuity and availability of navigation systems. They also benefit many other applications such as navigation, aviation, maritime, industry, and precision agriculture.

4.4 Applications in Digital Earth Case Studies

GNSSs have been widely applied in navigation for vehicles, offshore ships, aero craft and aerospace vehicles, geodesy, oil exploitation, precision agriculture, precise time transfer, Earth and atmospheric sciences, and many other fields. Its applications in the establishment of space-time geodetic datum, high-precision positioning and location-based services are introduced below.

4.4.1 Terrestrial Reference System

As a result of inexpensive GNSS receivers, densely distributed tracking stations and the high accuracy performance, GNSSs have played an important role in the establishment and maintenance of geodetic datum. The location and movement of a point on the Earth's surface must be expressed in a terrestrial reference system (TRS) attached to the Earth (also called the Earth-centered Earth-fixed system). The origin of the TRS is usually defined as the center of mass of the Earth, the Z axis is aligned with the international reference pole, the X axis is coincident with the Greenwich zero meridian, and the Y axis is orthogonal to the Z and X axes in the right-handed sense. As an ideal realization of the TRS, the TRF is comprised of a set of stations distributed on the Earth's surface with precisely known coordinates. The TRF is of great importance for geodesy, geophysics and space research. GNSS is an important data source to establish and maintain the TRF.

The widely used ITRF was established based on space-geodetic observations including GNSS, very long baseline interferometry (VLBI), satellite laser ranging (SLR), and Doppler orbit determination and radio positioning integrated on satellite (DORIS). As a realization of the International Terrestrial Reference System (ITRS), the ITRF can provide datum definitions (including origin, orientation and scale) for other global and regional TRS. Since 1988, more than ten versions of the ITRF have been released by the IERS, the latest version of which is the ITRF2014 released on January 2016 (Altamimi et al. 2016).

Different TRFs are adopted by different GNSSs. They include the World Geodetic System 1984 (WGS84) for GPS, Parametry Zemli 1990 (PZ-90) for GLONASS, Galileo terrestrial reference frame (GTRF) for Galileo and the BeiDou coordinate system (BDCS) for BDS. Most of the TRFs are aligned to the ITRF. The positioning results based on GNSS broadcast ephemeris are expressed in the corresponding TRF. As the TRF for GPS broadcast ephemeris, WGS84 has been refined several times by the US DoD, resulting in WGS84 (G730), WGS84 (G873), WGS84 (G1150), WGS84 (G1674) and WGS84 (G1762). PZ-90 is the TRF for GLONASS broadcast ephemeris. Successive versions of PZ-90, PZ-90.02 and PZ-90.11, have been released (Zueva et al 2014). GTRF is the TRF for the Galileo broadcast ephemeris, and GTRF07v01, GTRF08v01, GTRF09v01 and GTRF14v01 have been released (Gendt et al. 2011).

To unify the positioning results expressed in different TRFs, a 7-parameter Helmert transformation should be applied:

$$\begin{pmatrix} X_2 \\ Y_2 \\ Z_2 \end{pmatrix} = \begin{pmatrix} T_1 \\ T_2 \\ T_3 \end{pmatrix} + D \cdot R_1(\alpha_1) \cdot R_2(\alpha_2) \cdot R_3(\alpha_3) \begin{pmatrix} X_1 \\ Y_1 \\ Z_1 \end{pmatrix} \tag{4.2}$$

where T_1, T_2, T_3 are the translation parameters for the X, Y and Z axes, respectively, D is the scale factor, and $\alpha_1, \alpha_2, \alpha_3$ denote the Euler angles of rotation for the X, Y and Z axes. $R_i(\alpha_i)$ indicates the rotation matrix constituted by the rotation angles α_i for axis i, which can be expressed as:

$$R_1(\alpha_1) = \begin{pmatrix} 1 & 0 & 0 \\ 0 & \cos(\alpha_1) & -\sin(\alpha_1) \\ 0 & \sin(\alpha_1) & \cos(\alpha_1) \end{pmatrix}$$

$$R_2(\alpha_2) = \begin{pmatrix} \cos(\alpha_2) & 0 & \sin(\alpha_2) 0 \\ 0 & 1 & 0 \\ -\sin(\alpha_2) & 0 & \cos(\alpha_2) \end{pmatrix}$$

$$R_3(\alpha_3) = \begin{pmatrix} \cos(\alpha_3) & -\sin(\alpha_3) & 0 \\ \sin(\alpha_3) & \cos(\alpha_3) & 0 \\ 0 & 0 & 1 \end{pmatrix} \tag{4.3}$$

For the transformation parameters between different ITRF versions, please refer to Table 4.1 in the IERS Conventions (2010). The transformation parameters between ITRF2008 and WGS84, PZ-90, and GTRF versions are shown in Table 4.9. The definitions of the transformation parameters are the same as in Eq. (4.3).

4.4.2 Time System

Three types of time systems are commonly used. Their time scale are based on the Earth's rotation, e.g., the universal time (UT), the revolution of the Earth around the sun, and the electron transition frequency of atoms, e.g., the International Atomic Time (TAI). Coordinated Universal Time (UTC) uses the SI second of atomic time as its fundamental unit and is kept in time with the UT.

Atomic time is a time system based on the electromagnetic oscillation generated by the atomic transition inside substances. The Standard International (SI) second is defined as the time that elapses during 9,192,631,770 cycles of the radiation produced by the transition between two levels of the cesium 133 atom. The International Atomic Time (TAI) is the time system determined by the SI second, with the same origin as UT2 on 0 h 0 m 0 s, January 1, 1958. As a continuous and uniform time scale, TAI

Table 4.9 Transformation parameters between WGS84, PZ-90, GTRF and ITRF2008.

From	To	T_1 (mm)	T_2 (mm)	T_3 (mm)	D (ppb)	R_1 (mas)	R_2 (mas)	R_3 (mas)	Epoch
WGS84 (G1674)	ITRF2008	0	0	0	0	0	0	0	2005.0
PZ-90.11	ITRF2008	−3 ±2	−1 ±2	0 ±2	0 ±0.3	0.019 ±0.072	−0.042 ±0.073	0.002 ±0.090	2010.0
WGS84 (G1150)	PZ-90.02	360 ±100	−80 ±100	−180 ±100	0	0	0	0	2002.0
GTRF09v01	ITRF2008	2.0	0.9	4.7	−0.94	0.00	0.00	0.00	2000.0
Rate		−0.3	0.0	0.0	0.00	0.00	0.00	0.00	

is maintained by the Bureau International des Poids et Mesures (BIPM) using the atomic clocks of 400 national laboratories worldwide.

Universal Time (UT) is defined as the hour angle of mean sun relative to the Greenwich meridian plus 12 h. UT can be divided into three types. UT0 is directly determined from astronomical observations. Correcting the Earth's polar motion from UT0 yields UT1 whereas UT2 is obtained by correcting the seasonal variations of the Earth's rotation from UT1. UT1 defines the orientation of the average Greenwich meridian with respect to the mean equinox and thus represents the real rotation of the Earth. Since UT1 has a tendency of long-term slowdown, the difference between UT and the atomic time will grow increasingly larger. To avoid such an inconvenience, Coordinated Universal Time (UTC) has been adopted since 1972 based on the second length of TAI. It is a uniform but discontinuous time scale, and the difference between UTC and UT1, which is known as the leap second, is maintained within 0.9 s. The leap second with respect to UT1 is released by the IERS, and the relationship between TAI, UTC and leap seconds can be described as

$$TAI = UTC + leap \tag{4.4}$$

GNSSs are also an important technology to establish and maintain time systems. The GNSS time system is atomic time, in which the TAI second length is used and maintained jointly by high-accuracy atomic clocks onboard the GNSS satellites and implemented in the ground system. With an origin at 00:00 on January 1, 1980, the GPS Time (GPST) is adopted for GPS system. To ensure uniform continuity, there is no leap second in GPST and the constant difference between TAI and GPST is maintained as 19 s.

GLONASS makes use of GLONASS Time (GLONASST), which is synchronized to the UTC (SU) of Russia but biased by 3 h to match the local time zone of Moscow: GLONASST = UTC (SU) + 3 h. Unlike GPST, there are leap seconds in GLONASST.

Galileo adopts the Galileo System Time (GST) with an origin at 00:00 on August 22, 1999 (UTC time). The difference between GST and UTC at the starting epoch is 13 s and there is no leap second to maintain the uniform continuity of GST.

BDS makes use of BeiDou Time (BDT) with an origin at 00:00 on August 22, 1999 (UTC time). The difference between BDT and TAI at the origin moment is 33 s. BDT is counted with the week number (WN) and seconds of week (SOW). Similar to GST, no leap second is adopted to maintain uniform continuity of BDT. BDT is steered to UTC (NTSC).

The transformations between GPST, GLONASST, BDT, GST and UTC are as follows:

$$\begin{cases} UTC - GPST &= 19s - leap + C_0 \\ UTC - GLONASST &= C_1 \\ UCT - BDT &= 33s - leap + C_2 \\ UTC - GST &= 19s - leap + C_3 \end{cases} \tag{4.5}$$

where *leap* is the leap second of UTC with respect to TAI, as shown in Eq. (4.4); C_0, C_1, C_2, C_3 are the daily deviation values of GPST, GLONASST, BDT and GST relative to UTC, respectively, provided by the BIPM. The accuracies of C_0 and C_1 are approximately 10 ns and several hundreds of ns, respectively (ftp://ftp2.bipm.org/pub/tai/other-products/utcgnss/utc-gnss).

High-accuracy UTC time can be obtained through GNSS data. The GNSS common-view technique has been used by the BIPM for many years as one of its main techniques for international time transfer. It has the advantages of low equipment cost, high accuracy and convenient operation. In this technique, the time difference between two clocks, A and B, is determined by simultaneous observation of a third clock on a GNSS satellite. Each station observes the time difference between its clock and the GNSS time plus a propagation delay, which can be largely removed by using one-way GNSS time transfer procedures. By exchanging data files and performing a subtraction, the time difference between the two receiving stations is obtained.

The GNSS timing technique has been widely applied to time and frequency synchronization in the communication, finance and power industries in China. In the communication field, time synchronization for the whole communication network is realized through installation of GNSS timing terminals, so the billing time can be ensured to be consistent and accurate. For the frequency synchronization networks of China mobile, China telecom and China Unicom, the first-level reference clock and part of the second-level/third-level/micro-synchronization-node clock are equipped with a built-in GNSS reception module and external GNSS receivers. The time synchronization networks are also equipped with dual-mode GNSS timing receivers.

In the power industry, time and frequency synchronization for the substation network can be provided by GNSS timing. The time systems from power transmission network to power computer network in the Chinese power industries mainly use GPS as the master clock for timing and synchronization. On December 1, 2017, the 'Technical specification of time synchronization system and equipment for smart substation' (GB/T 33591-2017) became officially effective, in which the BDS is adopted as the main technique for time synchronization. By the end of 2017, there were nearly 900 sets of dispatching automation master station systems (in 11 categories) that could receive BDS signals in the domestic power grid control network, and more than 15,000 sets of GPS timing equipment have been updated to be compatible with the BDS.

4.4.3 High-Precision Positioning

The accuracy of single-point positioning based on broadcast ephemeris is only 10 m and is influenced by the unmodeled errors and noise of the pseudo-ranges. It cannot meet the requirements of many applications and limits the use of GNSSs. Differential GNSS techniques were developed to improve the positioning accuracy to decimeter-level. DGNSS/RTK and precise point positioning (PPP) are two commonly used high-precision differential positioning methods. The basic principle uses one or more reference stations with precisely known positions to model the observation errors, including the ionospheric and tropospheric delay and satellite clock and orbit errors to improve the accuracy and reliability of positioning for users.

The high-precision GNSS positioning algorithm was developed from the single-station pseudorange differential approach and carrier phase differential approach into a real-time carrier phase differential approach based on multiple reference stations (network RTK), PPP, and PPP with fixed ambiguity resolution (PPP-AR), improving the resolution accuracy and extending the application modes. The differential algorithms can be categorized into location differential, pseudorange differential and carrier phase differential techniques according to the differential observations adopted. The differential algorithms can also be classified as single-station differential, local area differential, wide area differential, or global real-time high-precision PPP based on the effective range of the differential corrections. They can be categorized into satellite-based and ground-based differential augmentation based on the type of broadcast link. Finally, the differential algorithms can be categorized into the state space representation (SSR) differential method and the observation space representation (OSR) differential method according to the differential model algorithm and parameters.

The PPP (SSR) and the network RTK (OSR) are the two major techniques in high-precision GNSS positioning services. The network RTK method, also known as the RTK method with multiple reference stations, usually needs more than three GNSS CORS stations within a certain region. Taking one or several stations as the reference stations, the distance dependent errors are modeled as regional OSR corrections. The differential corrections are provided to the rover stations in real time for precise positioning. The network RTK method can be classified into four types, including the virtual reference station (VRS) method, the master auxiliary concept (MAC) method, the Flächen Korrektur parameter (FKP) method or the combined bias interpolation (CBI) method, according to the OSR differential corrections used.

The VRS method is the most widely used network RTK technique at present. A virtual reference station is established near the rover station. Observation of the virtual reference station is generated using the real observation of the surrounding reference stations plus the regional error corrections. By receiving the observations of the virtual reference station, users can realize high-accuracy real-time positioning with the single-station RTK method. In the MAC method, corrections from the reference network can be divided into two categories: corrections closely correlated with the carrier frequency, e.g., the ionospheric delay, and corrections independent of the carrier frequency, e.g., the orbit error, tropospheric error, and multipath effect. The integer ambiguity of the reference network is initially fixed to ensure a uniform integer ambiguity reference for all the reference stations. The correction difference between the auxiliary station and the master station is calculated and broadcasted to the rover station. The principle of the FKP method is to estimate nondifferential parameters for each reference station in real time and generate the network solution. The spatial correlation error of the ionosphere and the geometric signal inside the network is then described with regional parameters. Based on these parameters and locations, the rover station computes the error corrections and realizes precise positioning.

The FKP method has been widely applied in Germany, the Netherlands and other European countries. The CBI method does not distinguish ionospheric delay from tropospheric delay and other types of errors when calculating the corrections of the reference stations. The corrections for each reference station are not broadcasted to the users. Instead, the observation data of all the reference stations are gathered to select, calculate and broadcast the comprehensive error corrections to the user.

For specific regional users, the accuracy of network RTK can achieve centimeter-level. However, due to the spatial restriction of OSR differential correction methods, the distance between reference stations in network RTK can generally be no more than 70 km. Therefore, it would be very costly to establish a wide-area real-time service system to serve a large number of users using the network RTK method. The PPP technique based on the wide-area (global) tracking network can realize high-accuracy positioning with only a few reference stations in a wide area. It could effectively overcome the disadvantages of network RTK. However, although PPP could provide positioning service with the same accuracy all over the world, it has the disadvantages of slightly lower positioning accuracy and relatively longer initialization time than network RTK.

Based on precise satellite orbit and clock error data, the PPP method could realize decimeter-level to millimeter-level positioning accuracy using carrier phase and pseudorange observations collected by a single GNSS receiver. Only the high-precision satellite orbit and clock errors are needed to obtain high-precision positioning for any station at any location and the positioning error is homogenous worldwide. Thus, PPP has been widely used in crustal deformation monitoring, precise orbit determination, precise timing, earthquake/tsunami monitoring and warning, and many other fields. As an extension of the standard PPP technique, PPP-AR can obtain ambiguity-fixed coordinates through restoring the integer characteristics of the nondifferential ambiguity. Its accuracy is equivalent to that of RTK.

In China, the first-generation BDS augmentation system was formally approved on April 28th, 1998, with the goal of providing GPS wide-area differential and integrity service for users based on BDS-1. It aims to improve the GPS accuracy and reduce the risk of using GPS. The first-generation BDS augmentation system (the first phase of construction) was completed and began trial operation in 2003. During this period, the augmentation system operated stably and provided real-time GPS differential correction and integrity service for various users in the service region. The positioning accuracy and integrity warning capability were basically in accordance with the design indicator requirements. In recent years, research and development of a wide-area real-time precise positioning prototype system in China and the neighboring areas have been carried out with the support of the national 863 program. As a key project in the field of Earth observation and Navigation Technique under the National High-tech Research and Development Program (863 program) in 2007, the 'wide-area and real-time precise positioning technique and prototype system' was jointly undertaken by the China Satellite Communications Corporation (China Satcom), China Center for Resources Satellite Data and Application, and Wuhan

University. Based on the wide-area differential and PPP technique, the satellite navigation augmentation service is realized with a positioning accuracy of better than 1 m for land, ocean and air transport in China.

The construction of CORS around world has entered into a new era. A provincial-level CORS system in Jiangsu and Guangdong provinces has been established in China. CORS systems have also been established in various large- and medium-sized cities, e.g., Beijing, Shanghai, Tianjin, Chongqing, Nanjing, Guangzhou, Shenzhen, Wuhan, Kunming, Jinan, Qingdao, Suzhou, Changzhou, Hefei, Dongguan, and Zibo. There are more than 2200 CORS stations in China. CORS systems may be upgraded to install BDS receivers. High-precision surveying can be conducted through these CORS systems with high efficiency and less man-power than traditional technology such as a total station. The CORS system is currently a vital part of surveying and mapping activities around the world, including urban planning, land surveying and mapping, cadastral management, urban and rural construction, and mining surveying.

The differential GNSS technique can support cadastral surveying to establish property boundaries, which is of great importance for fiscal policies such as land taxation. In the different construction stages of a building or civil engineering project (such as a highway, motorway, bridge, underground tunnel, railway, reservoir or embankment), GNSS positioning can be used to automatically control the construction equipment. GNSSs are also used to define specific location points of interest for cartographic, environmental and urban planning purposes. GNSSs play an important role in measurement and calculation at each stage of mine exploitation, including safety checks. GNSSs are used to monitor critical infrastructure and the natural environment to prevent major disasters and promptly intervene in case of emergency. GNSSs can support a wide range of activities in marine surveying, such as seabed exploration, tide and current estimation and offshore surveying.

4.4.4 Location-Based Service

Location-based service (LBS) systems work independently or cooperate with mobile terminals to provide real-time and post-processed positioning and timing service for various users through different communication networks. LBS relies on GNSS and augmentation systems to provide uniform space-time datum. Other assisted navigation and positioning techniques are also incorporated to improve the anti-jamming capability and availability of LBS. Through communication networks, e.g., the internet and mobile internet, LBS can provide users with positions, attitude, velocity and time synchronization services.

The workflow of a typical LBS system can be designed as follows: the GNSS wide-area augmentation system receives a real-time data stream from various GNSS tracking networks, generates the wide-area and regional satellite navigation augmentation signals, and provides them to the authorized public users through broadcasting systems controlled by a service provider. GNSSs are 'outdoor' positioning techniques, as the GNSS signal is affected by strong attenuation and multipath caused by complex indoor environments. In severe environments, the GNSS signals cannot be captured. Thus, the location of users inside a building should be determined by an indoor positioning system using WIFI, Bluetooth, INS, magnetic fields, and virtual beacons. The information integration platform receives the satellite navigation augmentation signals and merges them with geographical data to provide users with comprehensive location-based value-added service through LBS providers. As an integration of social networks, cloud computing and the mobile internet, LBS could become the core element of a series of significant applications, e.g., intelligent transportation systems (ITS), precise agriculture, intelligence manufacturing and smart cities. GNSS-enabled LBS applications are mainly supported by smartphones.

ITS refers to efforts to add information and communications technology to transport infrastructure and vehicles in an effort to manage factors that are typically at odds with each other, such as vehicles, loads, and routes, to improve safety and reduce vehicle wear, transportation times, and fuel consumption. GNSSs play an important role in ITS applications such as traffic control and parking guidance by providing accurate and reliable positioning. The low-cost high-precision GNSS receiver has a big potential market in ITS. The low-cost GNSS receiver can also be integrated with an inertial navigation system (INS) to develop an autonomous navigation system for general aviation (GA). General aviation is the term used to describe all aviation except government and scheduled-airline use.

The accuracy of GNSS SPS is only approximately 10 m. It cannot tell users the optimal lane to get to their destination, especially in dense urban environments such as multilane roads and highways. With the aid of an LBS system, lane-level navigation and positioning with meter-level accuracy can be realized. It will become the standard configuration for passenger vehicles and freight vehicles with hazardous chemicals in the future. The consortium within the EU-funded InLane project is working on the fusion between computer vision and GNSS technologies to achieve the required level of positioning that allows for the safe operation of autonomous vehicles (https://www.gsa.europa.eu/market/market-report).

The embedment of GNSS terminals in bicycle-sharing systems can result in more accurate and reliable positioning for better user experiences, especially in complex scenarios. The positioning accuracy can be improved from 50-100 m to approximately 3 m. The GNSS terminals can also support orderly parking. Currently, approximately half of the bicycle-sharing systems in China are equipped with GPS terminals. High-precision BDS positioning has also been adopted in driver training. It can automatically record the trail of the wheel at the centimeter level. Many driving test centers in China promote this technique.

The premise of precision agriculture is to adapt field operations to local variations in crop and soil conditions using state-of-the-art technology combined with knowledge-intensive field management. The positioning system is a part of precise equipment that consists of a differential global positioning system (DGPS) receiver, a radar velocity sensor, a wheel velocity sensor and an electronic compass. Precision positioning helps complete field applications faster and more productively, accurately, safely and comfortably, with less operator fatigue. GNSS is used in agriculture in a few key areas. As crops are harvested, a GNSS receiver connected to a yield monitor sensor records a coordinate along with the yield data. This data is combined and analyzed to create a map of how well different areas of the field are producing. When spreading fertilizer or planting, equipment operators have traditionally used markers such as foam or other visual aids to mark where they've been to try to avoid overlap. The assistance of GPS and onboard guidance systems such as a light bar, can further reduce overlap.

For many years, the leading technology for precision agriculture was GPS L1 receivers providing submeter precision. That precision can meet the requirements of applications at the submeter level, such as applying chemicals, field mapping and aerial spraying. However, high-precision applications such as auto-steer need centimeter precision. Historically, Hemisphere GPS (formerly CSI), Trimble Navigation, OmniSTAR, and smaller designers and system integrators have been the GNSS technology providers for precision agriculture. The world-wide agriculture market is booming. Auto-steer and other high-precision GNSS applications in agriculture have contributed to increased production capacity.

The GNSS navigation function in smart phones can record the wheel path and personal interests as well as the behaviors of pedestrians and drivers, providing large amount of social activity information. It should be regarded as an important source of big data on human activities and interests. In the future, with the application of high-accuracy navigation based on smart phones and the implementation of integrated indoor and outdoor location services, this big data will provide more abundant information. A 2013 Nature paper noted that the owner of a cellphone can be specified (with 95% probability) by analyzing the big data of the cellphone location tracks in a city with approximately 1,500 thousand people. LBS systems could also support applications such as geomarketing and advertising, fraud management and location-based billing, which require authentication of the position to protect app users.

LBS applications for healthcare are increasing. Healthcare needs are driving the diversification of wearables. For example, a GNSS-enabled haptic shoe allows for visually impaired users to set a destination in a smartphone app. The soles guide the user to the destination by vibrating in the front, back, or sides. Visually-impaired people or wheelchair users rely on a seamless navigation experience between outdoor and indoor environments. They need more high-precision horizontal and vertical position information (https://www.gsa.europa.eu/market/market-report).

In summary, there is a huge navigation and LBS market. The navigation and LBS network will also promote the development of industries such as national security, social security, energy conservation and emission reduction, disaster relief and mitigation, traffic and transportation, the IoT, resource investigation, and precision agriculture.

References

Altamimi Z, Rebischung P, Métivier L et al (2016) ITRF2014: A new release of the International Terrestrial Reference Frame modeling nonlinear station motions. Journal of Geophysical Research: Solid Earth 121(8): 6109–6131

BDS-ICD (2013) Space Interface Control Document. Open Service Signal. China Satellite Navigation Office. http://www.beidou.gov.cn/xt/gfxz/201812/P020181227529449178798.pdf. Accessed 13 December 2018

Gendt G, Altamimi Z, Dach R et al (2011) GGSP: realization and maintenance of the Galileo terrestrial reference frame. Advances in Space Research 47(2):174–185

Grimes JG (2007) Global positioning system precision positioning service performance standard. GPS Navster. Department of Defense. https://www.gps.gov/technical/ps/2007-PPS-performance-standard.pdf. Accessed 13 July 2019

Gurtner W, Estey U (2013). The Receiver Independent Exchange Format. http://nebula.wsimg.com/5d20c8382d99be2caba8884b018c8dbf?AccessKeyId=2D3610E97C8E55B553AE&disposition=0&alloworigin=1. Accessed 13 July 2019

ICD-Galileo (2008) Galileo open service, signal in space interface control document (OS SIS ICD). Edited by European space agency/European GNSS supervisory authority. https://www.gsc-europa.eu/system/files/galileo_documents/Galileo-OS-SIS-ICD.pdf. Accessed 13 July 2019

ICD-GLONASS (2008) Glonass interface control document. Russian Institute of Space Device Engineering, Moscow, Russia. https://www.unavco.org/help/glossary/docs/ICD_GLONASS_5.1_(2008)_en.pdf. Accessed 13 July 2019

ICD-GPS-200J (2018) Space Segment/Navigation User Interface Control Documents. https://www.gps.gov/technical/icwg/IS-GPS-200J.pdf. Accessed 13 July 2019

Zueva A, Novikov E, Pleshakov D et al (2014) System of Geodetic parameters. Parametry Zemli 1990, (PZ-90.11). Paper presented at UN International Committee on Global Navigation Satellite Systems. Working Group D: Reference Frames, Timing and Applications. Working Group Meeting

Chuang Shi is a professor at School of Electronic and Information Engineering, Beihang University. He graduated from Wuhan Institute of Surveying and Mapping (current Wuhan University) and obtained his Ph.D. in 1998. His research interests include network adjustment, precise orbit determination of GNSS satellites.

Na Wei is an associate professor at GNSS Research Center of Wuhan University. She has obtained her Ph.D. degree in 2011 from Wuhan University. Her research interests include terrestrial reference frame and surface mass loading.

5

GIIs, GIS and SDI

Sven Schade, Carlos Granell, Glenn Vancauwenberghe, Carsten Keßler, Danny Vandenbroucke, Ian Masser and Michael Gould

Abstract Geospatial information infrastructures (GIIs) provide the technological, semantic, organizational and legal structure that allow for the discovery, sharing, and use of geospatial information (GI). In this chapter, we introduce the overall concept and surrounding notions such as geographic information systems (GIS) and spatial data infrastructures (SDI). We outline the history of GIIs in terms of the organizational and technological developments as well as the current state-of-art, and reflect on some of the central challenges and possible future trajectories. We focus on the tension between increased needs for standardization and the ever-accelerating technological changes. We conclude that GIIs evolved as a strong underpinning contribution to implementation of the Digital Earth vision. In the future, these infrastructures are challenged to become flexible and robust enough to absorb and embrace technological transformations and the accompanying societal and organizational implications. With this contribution, we present the reader a comprehensive overview of the field and a solid basis for reflections about future developments.

Keywords Geospatial information · Infrastructure · Spatial data · Public sector · Government

---

S. Schade (✉)
European Commission—Joint Research Centre, Ispra, Italy
e-mail: s.schade@ec.europa.eu

C. Granell · M. Gould
Universitat Jaume I, Castellón, Spain

G. Vancauwenberghe · D. Vandenbroucke · I. Masser
Katholieke Universiteit Leuven, Louvain, Belgium

C. Keßler
Aalborg University, Copenhagen, Denmark

M. Gould
ESRI, Redlands, USA

5.1 Introduction

Geospatial information (GI), i.e., information including a relationship to the Earth (Worboys and Duckham 2004), is a foundational ingredient for any Digital Earth application. Examples include information about land parcels, transport networks and administrative boundaries, vehicles, microplastics, fine particles, mobile devices and people. With GI, we can build digital replicas of our planet and use them to exchange knowledge, monitor the state of the Earth, simulate possible future scenarios or assess possible impacts of decision making. Although also other terms (such as 'geographic' or 'spatial') are used in scientific and other literature to refer to the same or similar concepts, we use 'geospatial' in this chapter. Furthermore, we speak of 'information' as (possibly processed) data in a context that allows for interpretation and meaningful use.

The technological, semantic, organizational and legal structure that allows for the discovery, sharing, and use of GI is called geospatial information infrastructure (GII) (Yang et al. 2010; Granell et al. 2014). With its core functionalities, a GII can be considered the backbone for Digital Earth. GIIs are essential to facilitate the information flow that is required to implement any past, present and future version of the Digital Earth vision—the knowledge sharing platform as initially envisaged by Gore (1998), a global tool for multidisciplinary research as outlined by Goodchild and colleagues (2012), or the world laboratory to support codesign, cocreation and codelivery that was suggested by Schade and Granell (2014).

Emerging from an initially highly technical concept, GIIs have a relatively long history and are well researched, including their close relationships with geographic information systems (GIS) (Worboys and Duckham 2004) and spatial data infrastructure (SDI) as enabling technologies (Masser 2005; Yang et al. 2010). Prominent examples of these enabling technologies include the spatial data infrastructure for Australia and New Zealand (ANZLIC 2019), the United States of America-US National Spatial Data Infrastructure (NSDI 2019), the Infrastructure for Spatial Information in the European Community [(INSPIRE 2019), see also Chap. 20], OpenStreetMap [(OSM 2019a), see also Chap. 18], and Google Maps (Google 2019).

By nature, GIIs undergo a continuous evolution that is primarily driven by the increasing pace of technological advancements and the inherent digital transformation of our societies (Castells and Cardoso 2005; Gimpel and Röglinger 2015). Similar to other information handling tools, GIIs face continuous challenges caused by the speed of technological progress that sometimes conflicts with the heaviness inhering in most governance structures. For example, the implementation of heavily governed GIIs bears a risk to continually run behind technological solutions (Schade and Smits 2012; Tsinaraki and Schade 2016). We have witnessed a shift from public sector (alone) to more collaborative approaches to the provision and operation of GI and related services, which increasingly involve the private sector (smeSpire 2014; Sjoukema et al. 2017). Whereas public sector information (e.g., about cadastral parcels or protected sites) continues to play an important role, increasing amounts of spatial data are produced, owned and provided by the private sector. Examples include street (navigation) data and satellite imagery, and 'standard' products such as Google Earth, Google Maps, Bing, and spatial data from GPS providers such as Here and TomTom.

In this situation, we face two opposing forces: traditional standardization processes and frequent technological disruption. Heavily standardized large infrastructures and platforms to support Digital Earth may have been a necessity a few years ago (Granell et al. 2016), when large amounts of GI were not easily accessible and data transformation used to be a process that was run on large computing machines for a long time before harmonized information could be provided to users. During that time, it was affordable to invest in traditional standard-based infrastructures and in educational programs that provided specialized training to develop, maintain and use such infrastructures (Masser 2005; Vandenbroucke and Vancauwenberghe 2016). However, is this still affordable today—in an era of fast digital transformation when disruptive technologies are about to become a new norm? Or will microservice-based architectures (Dragoni et al. 2017) to build smaller, more manageable platforms beat monolithic, big, layered architectures? How must the development of standards change to fit these new dynamics? What roles will the private sector play in this new set-up?

The question of whether Digital Earth will follow the traditional standardization approach, an alternative approach that completely embraces vivid digital transformation or anything in between has strong implications on the definition of the conceptual architecture of the GII with Digital Earth. Hence, we are at a controversial point in GII and Digital Earth history. This chapter outlines how we arrived at this point, explains the current situation in more detail, provides a critical reflection, and outlines a few future trajectories. We hope that this contribution to the Manual of Digital Earth aids in understanding the importance and evolution of GIIs and provides food for thought for those that will develop and use GIIs to implement the Digital Earth vision.

The remainder of this chapter is structured as follows. The next section introduces the history of GIIs during the different phases of organizational and technological development (Sect. 5.2). Next, we outline the current situation in respect to GII development, education and use (Sect. 5.3). We focus on the evolving relationship between GIIs and Digital Earth and important recent movements such as Open Science. In Sect. 5.4, we discuss changes and the challenges that GIIs face today. The most important implications for the Digital Earth vision are highlighted. In Sect. 5.5, we close the chapter with a brief conclusion and an outlook on the future of GIIs in support of Digital Earth. For details about GI analysis and processing, we refer the reader to Chap. 6. Matters of GI visualization are discussed in Chap. 7.

5.2 A Brief History of Geospatial Information Infrastructures

GIIs are not a new concept, and have evolved over a series of generations, each characterized by changing purposes, available technologies, and the main stakeholders involved in their design, implementation and use. Instead of describing these generations in detail, which has been done elsewhere (Rajabifard et al. 2002; Yang et al. 2010), we highlight fundamental milestones in the history of GII. We also highlight evolutions of the technical architectures used to implement GIIs over the past few decades.

5.2.1 Geospatial Information Infrastructure Milestones

In the history of GIIs worldwide, a series of milestones have been essential for the evolution of GIIs into their current form—most of which relate to actions of government, i.e., policy updates. Notably, these milestones differ in nature, for example, by administrative dimension, research purpose or geographic extent. However, they give a sensible impression of aspects that have framed the evolution of GIIs up to today.

As a first milestone, the EU initiated the CORINE program in 1985 with the aim of describing the status of the environment in Europe. This program was the first large-scale effort in Europe to collect spatial data covering the European territory according to agreed specifications in view of supporting different policies. It delivered its first pan-European land cover data set in 1990, with updates in 2000, 2006 and 2012. The second milestone dates back more than thirty years to the establishment of the Australian Land Information Council (ALIC) in January 1986. ALIC was the result of an agreement between the Australian Prime Minister and the heads of the state governments to coordinate the collection and transfer of land-related information between the different levels of government and to promote the use of that information in decision making (ANZLIC 1992). One year later, a third milestone occurred in May 1987 with the publication of the Report of the British Government Committee of Enquiry on Handling Geographic Information chaired by Lord Chorley (Coppock 1987). This report, also known as the Chorley report, set the scene for much of the subsequent discussion about GIIs in the UK and in other parts of the world. While the report reflected the committee's enthusiasm for the new technology: "the biggest step forward in the handling of geographic information since the invention of the map" (para 1.7), it also expressed their concern that information technology must be regarded as "a necessary, though not sufficient condition for the take up of geographic information systems to increase rapidly" (para 1.22). A fourth important milestone in the late 1980s was the release of the first issue of the International Journal of Geographic Information Systems, also in 1987. The journal, renamed the International Journal of Geographic Information Science in 1997, was the first scholarly journal devoted to GI.

The fifth milestone occurred in 1990 when the United States Office of Management and Budget (OMB) established an interagency Federal Geographic Data Committee (FGDC) to coordinate the "development, use, sharing, and dissemination of surveying, mapping, and related spatial data." The main objectives of a national GII were "encouraging the development and implementation of standards, exchange formats, specifications, procedures, and guidelines, promoting technology development, transfer, and exchange; and promoting interaction with other existing Federal coordinating mechanisms that have an interest in the generation, collection, use and transfer of spatial data..." (OMB 1990, pp. 6–7). These ideas were subsequently developed and extended by the United States National Research Council's Mapping Science Committee in their report 'Toward a coordinated spatial data infrastructure for the nation' (National Research Council et al. 1993). This report, which can be seen as a sixth milestone in the history of GIIs, recommended that effective national policies, strategies, and organizational structures be established at the federal level for integration of national geospatial data collection, use and distribution. A seventh milestone is the outcome of an enquiry by the Directorate-General XIII (now DG Connect) of the European Commission (EC), which found that there was a strong Europe-wide demand for an organization that would further the interests of the European GI community. As a result, the first regional level multidisciplinary SDI organization in the world was set up in 1993. The vision of the European Umbrella Organisation for Geographic Information (EUROGI) was not to "replace existing organisations but catalyse effective cooperation between existing national, international, and discipline-oriented bodies to bring added value in the areas of Strategy, Coordination, and Services" (Burrough et al. 1993).

An eighth milestone that marks a turning point in the evolution of the SDI concept came in the following year with the publication of Executive Order 12906 signed by President Bill Clinton, entitled "Coordinating Geographic Data Acquisition and Access: the National Spatial Data Infrastructure" (Executive Office of the President 1994). This described the main tasks to be carried out and defined time limits for each of the initial stages of the national spatial data infrastructure. These included the establishment of a national geospatial data clearing house and the creation of a national digital geospatial data framework. (Here, we understand data clearing houses as "internet-based components that intend to facilitate access to spatial data, by establishing a centralized site from which data from several sources can be found, and by providing complementary services, including searching, viewing, transferring, and ordering spatial data" (Davis 2009). The Executive Order gave the FGDC the task of coordinating the Federal government's development of the National Spatial Data Infrastructure. As the Executive Order also required each member agency of that committee to hold a policy-level position in their organization, it significantly raised the political visibility of geospatial data collection, management and use among US institutions and internationally. The organization of the first Global Spatial Data Infrastructure (GSDI) Conference in Bonn, Germany, in September 1996 was another—ninth—milestone in the 90s. The conference brought together representatives from the public and private sectors and academia for the first time to discuss matters relating to NSDIs at the global level. Shortly after, in 1998, the Baveno Manifesto set a fundamental milestone for European space policy. It led to the establishment of the Global Monitoring for Environmental Security (GMES) program, which was formally established in 2010 (Regulation (EU) No 911/2010), and followed by the Copernicus program in 2014 (Regulation (EU) No 377/2014).

After 2000, the evolution of GIIs worldwide continued, and several milestones can be highlighted. One was the establishment of the intergovernmental Group on Earth Observations in February 2005 to implement a global Earth observation system of systems (GEOSS) to integrate observing systems and share data by connecting existing infrastructures using common standards. In 2018, there were more than 400 million open data resources in GEOSS from more than 150 national and regional providers such as NASA and ESA, international organizations such as the World Meteorological Organization (WMO) and commercial sector groups such as Digital Globe (Nativi et al. 2013). Another—eleventh—milestone was the launch of the first scholarly journal in the GII field in 2006. The International Journal of SDI Research is a peer-reviewed journal that is operated by the Joint Research Centre of the European Commission, which aims to further the scientific endeavor underpinning the development, implementation and use of Spatial Data Infrastructures. Directive 2007/2/EC of the European Parliament and of the Council of 14 March 2007 established an Infrastructure for Spatial Information in the European Community (INSPIRE, see Chap. 20 for more details) can be seen as a twelfth milestone in the evolution of GIIs. The INSPIRE Directive aimed to establish a spatial data infrastructure to improve the sharing and interoperability of geospatial data in support of environmental policies and policies that might have an impact on the environment (Directive 2007/2/EC of the European Parliament and the EU Council) and was the second multinational GII initiative that sought to make harmonized high-quality GI readily available. INSPIRE stresses the principles of data sharing and cross-border usage of the data. The year 2011 marked a key event that initiated the deep involvement of the private sector: the first Geospatial World Forum "Technology for people and Earth" (Geospatial World 2011). This global conference gathers diverse stakeholders to present and discuss the pathways of the geospatial industry. In addition, the United Nations Committee of Experts on Global Geospatial Information Management (UN-GGIM) was established in July 2011 (ECOSOC Resolution 2011/24) as the official United Nations consultative mechanism on global GI management. Its primary objectives are to provide a forum for coordination and dialogue among Member States of the United Nations (UN) and between member states and relevant international organizations and to propose work plans that promote global frameworks, common principles, policies, guidelines and standards for the interoperability and interchangeability of geospatial data and services. Not long ago, (23 June 2015) the first Sentinel (satellite developed for the Copernicus program delivering open Earth Observation) was launched. This milestone initiated the launch of a set of sister satellites that deliver high-resolution images and contribute strongly to a new era of GI provision worldwide.

Table 5.1 provides an overview of the fifteen milestones discussed in this section. They are mostly institutional, legal and policy-related. Notably, the milestones cover different regions (e.g., Europe, North America and Australia), administrative levels (e.g., national, regional and global) and sectors (e.g., academic and cross-sectoral initiatives). They reflect the breadth and diversity of GII initiatives since the 1980s. This demonstrates how GIIs took a leading role in promoting and enabling open data publishing, possibly as the most common theme across the globe. These developments took place in support of the Open Movement, Open Science, Open GIScience, and citizen science, which we explore in more detail later in this chapter (in Sect. 5.3).

Table 5.1 Geographical information infrastructure milestones

Year	Milestone
1985	The European Union (EU) launched the CORINE land cover program as the first large-scale effort to collect spatial data covering the European territory
1986	The Australian Land Information Council began coordinating the collection and transfer of land-related information between the different levels of government
1987	Report of the Committee of Enquiry into Handling Geographic Information, chaired by Lord Chorley
1987	Launch of the International Journal of Geographic Information Systems
1990	The US Federal Geographic Data Committee was created to coordinate the development, use, sharing and dissemination of surveying mapping and related geospatial data
1993	US Mapping Science Committee report on 'Toward a coordinated spatial data infrastructure for the nation'
1993	Establishment of the European Umbrella Organisation for Geographic Information (EUROGI) as the first regional-level multidisciplinary SDI organization
1994	Executive Order 12906 'Coordinating geographic data acquisition and access: the National Spatial Data Infrastructure'
1996	First Global Spatial Data Infrastructure conference in Bonn, Germany
2005	Establishment of the intergovernmental Group on Earth Observations in February 2005 to implement a Global Earth Observation System of Systems (GEOSS)
2006	Launch of the International Journal of Spatial Data Infrastructure Research
2007	Directive 2007/2/EC of the European Parliament and of the Council of 14 March 2007 establishing an Infrastructure for Spatial Information in the European Community (INSPIRE)
2011	The first Geospatial World Forum 'Technology for people and Earth' took place
2011	The United Nations Committee of Experts on Global Geospatial Information Management (UN-GGIM) was established
2015	Launch of Sentinel-2, the first of a series of Copernicus satellites delivering open and high-resolution GI

5.2.2 *Architectural Evolutions in Geospatial Information Infrastructure Development*

Alongside these milestones, we have witnessed an evolution of GII architectures and technological solutions, following the increased sophistication of technology and growth in user requirements. We summarize the central developments, concentrating primarily on GII. Another chapter of this manual addresses the irruption of sensors, sensor networks and the Internet of Things (IoT, see also Chap. 11). For developments and implications of machine learning, deep learning, and artificial intelligence, we refer the interested reader to Chap. 10.

GI has been used for many decades in different application fields (Longley et al. 2011). In the eighties, GIS technology started to spread globally. Prior to the development of SDIs, which only expanded at a broader scale in the nineties, geospatial data assets were created, managed and used by individual organizations using standalone GIS. In 1987, Specialized software companies such as ESRI brought GIS software to the market that could run on personal computers. Others, such as Intergraph, did the same, and academic and even public sector parties developed software for using geospatial data for particular purposes. Examples include ILWIS (2019) and GRASS (2019).

However, in this period, most efforts were focused on the collection and maintenance of data, as well as its use within the organization. Big data collection efforts started taking place. For example, in Europe the need for data that are standardized, well-documented, high-quality and available for the broader community became apparent. Therefore, the Coordination of Information on the Environment—CORINE program was initiated by the EU. This and similar initiatives elsewhere in the world, e.g., in the US through the FGDC, revealed the need to work more systematically in several technical aspects: documentation (metadata), data harmonization, access mechanisms and standards (Nebert 2004). These technological developments occurred in parallel with the organizational and institutional developments described in the previous section.

Originally, the focus was on exchange formats and particularly on the transformation—where required—from one format to another. In practice, for a long time de facto standards were used a lot. One good example is the shapefile format developed by ESRI that was (and still is) used to transfer geospatial data from one organization to the other (ESRI 1998). In the nineties, data exchange between organizations, although often on an ad hoc basis, became more and more important to avoid duplication of data sets and to share resources more efficiently. With this increased exchange, good documentation became paramount (Danko 2005). From the early nineties, organizations explored ways of documenting data in a standard manner. The first standard used by many was the FGDC metadata standard, which was initiated by President Clinton by executive order 12906 (1994) and became official in 1998 (FGDC 1998). This standard was used a lot, even in Europe. Work on an international metadata standard also began in the nineties but saw only light in 2003 with the adoption and publication of the International Organization for Standardization (ISO) standard

19115 in 2003 (ISO 2003). Since then, thousands of organizations have documented their geospatial data sets according to this standard.

To make potential interested parties aware of the existence of geospatial data resources, the publish-find-bind paradigm was defined as a key concept for SDIs (van Oosterom 2005). The idea is to 'publish' geospatial data resources by documenting and putting them in a catalogue, then make them 'discoverable' by a search mechanism and 'accessible' through a binding mechanism, which means that they can be integrated in a user application (e.g., a web-viewer, a desktop GIS or other application). In addition to the metadata, access mechanisms were designed and developed and became a key component of the technical parts of an SDI (Zhao and Di 2011). The Open Geospatial Consortium (OGC), which was established in 1994, brought together different academic, private and public sector parties and soon focused on the standardization of interfaces for accessing geospatial data resources (OGC 2019). They developed several web service interfaces to perform basic jobs such as 'discovery' (CSW—Catalogue Services for the Web), 'viewing' (WMS—Web Mapping Services—a first version of the standard was released in 2000), and 'downloading' (WFS—Web Feature Services). These OGC standards were meant to adhere to Service-Oriented architecture (SOA), an architectural style aimed at designing applications based on a collection of best practices, principles, interfaces, and patterns related to the central concept of a service (Papazoglou and van den Heuvel 2007). In SOA, services are the basic computing unit to support development and composition of larger, more complex services, which can be used to create flexible, ad hoc, dynamic applications. The main design principle behind SOA is that a service is a standards-based, loosely coupled unit composed of a service interface and a service implementation. Previous examples of OGC service specifications were designed to comply with these SOA principles to publish, find and use geospatial data. Most of the commercial software companies, as well as the Free and Open Source Software for Geospatial (FOSS4G) community—with a major push in 2006 with the establishment of the Open Source Geospatial Foundation (OSGeo)—developed tools and platforms to create such services and build portals for users to find and access data (Tait 2005; Maguire and Longley 2005).

In addition to the efforts to document GI and make it more discoverable and accessible, many efforts were made to better harmonize them for use in cross-border and cross-sector settings. The ISO Technical Committee 211 (ISO/TC211) was created in 1994 to look into the standards for Geographic Information and Geomatics. There was a large effort to develop the so-called ISO 19100 series of standards that, in addition to the already mentioned metadata standard, comprise a series of standards describing how to model our world (ISO 2002). The series includes a reference model, the definition of spatial and temporal schema, rules for application schema, and a methodology for cataloguing spatial features. In 2001, preparations began to design and implement INSPIRE, the Infrastructure for Spatial Information in Europe (INSPIRE 2019). In addition to the key idea of improving GI sharing (policy challenge), another objective was to improve spatial data interoperability through the design of data specifications for 34 themes (technical and organizational challenges). The ISO 19100 series of standards served as a basis for this huge effort. The resulting portfolio of standardized data sets serves cross-border applications in the context of environmental policy making/monitoring and other sectors. The process of harmonization is still ongoing at the time of writing.

In parallel with these more formal developments, stimulated by the public sector, developments in the private sector soon influenced SDIs and were ultimately (at least partially) integrated. In 1998, US Vice President Al Gore coined the term Digital Earth in view of global challenges such as climate change (Gore 1998). In 2001, a small software company called Keyhole Inc. launched and developed the Keyhole Earth viewer for looking at our globe from a global, bird's-eye view. The technical solution, aimed primarily at the defense sector, looked at geospatial data from a 3D perspective. A few years later, in 2004, Google acquired the small company and launched the still very popular product Google Earth based on the KML standard. More SDI developments embraced these new developments and aimed to integrate data from these commercial products into SDIs and applications. This whole process ended with the adoption of KML (originally keyhole markup language) by the OGC as a community standard.

In support of the development of SDIs, specific software developments emerged. Traditional GIS desktop software such as ArcGIS from ESRI was extended with server and mobile software, and the FOSS4G communities developed specific products that became very popular, such as GeoNetwork (to create geoportals and catalogue services), GeoServer, Degree and others (to set up all kinds of web services), and open source systems for data management such as POSTGIS (Steiniger and Bocher 2009; Brovelli et al. 2017). In addition to open standards and open software, open data became a new paradigm with important initiatives that are still very popular. In 2004, there was a global effort from the geospatial community to develop and maintain a network of streets (OSM), which was a joint effort of thousands of volunteers to provide and include data into an open data product (Haklay and Weber 2008). These volunteered geographic information (VGI) efforts became also part of the maintenance procedures of commercial products such as those from TomTom (formerly Tele Atlas) and Here (formerly NavTeq). The idea of citizens contributing to data and information gathering has now become widespread and is termed crowdsourcing [(Capineri et al. 2016), see also Chap. 18 for more details]. From the SDI perspective, which often focuses on authoritative data coming from government, these initiatives and the resulting geospatial data resources are considered complementary. The concept of GIS-based (open) data portals for smart city projects has also taken hold in recent years (ESRI 2019a).

These developments (see Table 5.2) led to a vibrant geospatial community and many GIIs that are interconnected (Vandenbroucke et al. 2009) and rich in content and quality, and new developments started to influence the way of working and the methods of providing data to user communities. Although the geospatial world has always worked somewhat in isolation, the developments in the general ICT world led to increased interest in joining forces. In 2006, Tim Berners-Lee coined the concept of Linked Data to combine the huge amounts of data available on the web (Berners-Lee 2006). In the geospatial world, this led to the idea of the geosemantic web. In 2014, the OGC and W3C started several joint initiatives including the Spatial Data

Table 5.2 Timeline of relevant technological developments

Year	Technological development
1982	First release of GRASS as open source software, managed by the US Army Corps of Engineers
1985	First release of ILWIS as closed, proprietary software developed by a university, ITC in the context of a land use zoning and watershed management project in Sumatra; the release as open source software followed in 2002
1987	First release of pcARC/INFO (ESRI), available on personal computers
1992	Introduction of the shapefile format by ESRI, which became a de facto standard format for exchanging geospatial data
1994	The Open Geospatial Consortium (OGC) is established and starts work with 8 members (currently more than 500)
1994	ISO/TC 211 is created as one of the technical committees of the International Organisation for Standardisation (ISO) responsible for the field of Geographic Information
1998	The FGDC metadata standard (US) becomes official
1998	The term Digital Earth is coined by former US Vice President Al Gore, describing a virtual georeferenced representation of the Earth
1998	Publication of the specifications of the shapefile format, which became open at that stage
2000	First version of the Web Mapping Service (WMS) interface standard released by OGC
2001	Keyhole Inc., the developers of Google Earth (originally called Keyhole Earth Viewer) and the KML format, is established
2002	The development of quantum GIS (QGIS) began, released as open source in 2009
2003	ISO 19115 Geographic Information—Metadata standard is adopted by the participating countries
2004	Publication of the SDI Cookbook by the Global Spatial Data Infrastructure Association (GSDI)
2004	Keyhole Inc. is acquired by Google
2004	OSM launched
2006	Linked Data as a concept, method and technique was coined by Tim Berners-Lee within the W3C as part of the semantic web project
2006	Founding of the open source geospatial foundation OSGeo, with currently more than 30000 volunteers, and the first FOSS4G International Conference in Lausanne, Switzerland
2007	The term volunteered geographic information (VGI) was coined by Michael Goodchild
2008	INSPIRE metadata regulation adopted as Implementing Rule 2007/2/EC
2010	INSPIRE regulation regarding interoperability of spatial data sets and services adopted as Implementing Rule 1089/2010
2014	The DCAT metadata standard for data resources of W3C is released
2014	Establishment of the Spatial Data on the Web Working Group focused specifically on the intersection of issues facing OGC and W3C members
2015	OGC adopts KML as a community standard
2015	GeoDCAT-AP, an implementation allowing for data exchange between geoportals and open data portals, was adopted by the EU ISA program

on Web Working Group to examine and test new ways of publishing and linking data (W3C 2015). One of the tangible results of this closer collaboration was the effort to exchange metadata between geoportals and (open) data portals through a more generic and broadly used DCAT standard (W3C 2014). More of these developments are expected to take place—and will continue to emerge faster and faster. This poses particular challenges to standardization processes, which should keep pace with these evolutions. In the next sections, we describe ongoing and new developments, for example, the changing power relationship from the public to the private sector and challenges posed by the trending platformization of society (van Dijck et al. 2018).

5.3 Geospatial Information Infrastructures Today

Leaving the past behind, several important developments and aspects of the current situation of GIIs are important to mention. We consider the following items worth highlighting in the context of Digital Earth: the mainstreaming of GI and the proliferation of GIIs, especially on the web; the contribution of GII developments to the opening of data and science as a whole; and the growth of knowledge exchange and learning networks across the globe.

5.3.1 The Evolution of Geospatial Information on the Web

In parallel to the organizational milestones described in Sect. 5.2, the technical foundations of GIIs were developed and standardized, mostly through bodies such as the OGC and ISO (see Sect. 5.2.2). Along with this development, the proliferation of slippy web maps and map-based mobile applications has led to the establishment of a separate branch of GII that primarily addresses end-user needs by, for example, providing directions to get from one place to another or offering extensive geocoding capabilities ("where is the closest coffee shop?"). The widespread adoption of these new services that were no longer just providing GI for a group of professional users was driven by the introduction of Google Maps in 2005 as well as the introduction of touch-screen smart phones with built-in GPS through the first iPhone in 2007.

Shortly after the introduction of Google Maps, the first reverse-engineered map mashups appeared and demonstrated how Google's JavaScript-based maps could be combined with GIs from other sources (such as crime data on ChicagoCrime.org or real estate offerings on housingmaps.com). The subsequent release of a public Application Programming Interface (API) for Google Maps that allowed for any web developer to embed a map in their web pages triggered the development of open source alternatives such as OpenLayers (OpenLayers 2019), which is still under active development today as an OSGeo project. OpenLayers is notable in this context because it bridges the worlds of consumer-oriented GI and GIIs targeted at professional use cases by allowing the combination of data from OGC-based web services with tiles and file formats such as KML. Esri released its JavaScript API to facilitate the creation of web apps from traditional GIS datasets (ESRI 2019b). Together with other, more recent examples such as Leaflet (Leaflet 2019), a 'grass-roots' standard emerged for the URL scheme of map tiles for slippy maps (OSM 2019b). This URL scheme enabled any web mapping framework to consume and display the tiles from any of the increasing number of servers that can produce them (GeoServer, MapServer and TileStache are a few examples)—a de facto standardization process that was successfully completed without the involvement of any of the abovementioned standardization bodies.

A third aspect that explains today's GII landscape is the development and widespread adoption of open data (Open Data Barometer 2015; Gurstein 2011; Kitchin 2014). This includes thematic open data sets available for direct download via web portals (ESRI 2019c) and the free provisioning of governmental data, which was previously made available for a (sometimes substantial) fee—if at all. The proliferation of open government data (OGD) has been complemented by the development of user-generated data sources, dubbed volunteered geographic information (VGI, Goodchild 2007) in the context of GI. OGD aims to make data originally produced for professional users available to a broader public, and VGI can be seen as a grass-roots movement producing its own collection of non-authoritative datasets. The VGI project with the most profound impact is OSM. OSM started as a free, bottom-up alternative to the then-prohibitively expensive data produced by the UK's Ordnance Survey, and has since become the largest collection of freely available GI. At the time of writing of this chapter, the OSM database consisted of close to 5 billion mapped nodes (points), collected by almost 5 million registered users (OSM 2019c). Its significance for development in the field today lies in the provisioning of a free, global, and in many areas extremely detailed collection of GI and in the number of innovative companies that have entered the market with products based on OSM. They continuously contribute to the OSM dataset and have developed open source tools around it for mapping, quality checking, and the use and processing of OSM data.

A notable recent development that emerged from this OSM ecosystem is the trend towards using vector tiles instead of prerendered image tiles. The improved support for rendering vector data in modern web browsers has enabled this switch. Vector tiles have several advantages over image tiles, such as adaptable styling, maps that look sharp independent of screen resolution, and opportunities for interaction with the actual individual map features (interactive labels or clickable features, for example), with smaller data volumes to transfer between the server and client.

As these examples show, the collection, distribution, and analysis of GI has evolved from a field that used to require expensive equipment and extensive professional training to activities that are carried out by users (and contributors) with highly diverse backgrounds and different levels of education. The ubiquity of devices capable of both producing and consuming GI, in combination with an ever-growing amount of free-to-use GI and powerful free and open source software solutions has led to a somewhat chaotic landscape of practices, standards and conventions for the data and processes involved. The involvement of a broader public in these processes also means that organizations such as OGC or ISO that primarily focus on the professional use of GI are addressing a decreasing share of the actual GI user base. An increasing number of users and producers of GI do not work for government agencies, conduct commercial mapping efforts, or develop software for GI web services. Instead, they may be working in data science or data visualization (Bostock et al. 2011), may be open data advocates or citizen scientists, or may do research in areas such as economics, ecology, or the humanities.

New companies that deal with GI at the core of their business that do not consider themselves GIS companies have established new ways of dealing with GI, without taking the time to go through time-consuming standardization processes. The long list of prominent examples of these companies includes Mapbox, Carto, Uber, booking.com, Trip Advisor, Google and Facebook. Many of the relatively new internet platforms contain GI and thereby initiated a shift to the traditional organizational structures (van Dijck et al. 2018).

Arguably, with this industrial production and use of GI, large companies set the de facto standards—as far as standards are relevant for their internal workings. To some extent, these developments have been acknowledged by the World Wide Web Consortium (W3C) in some efforts that have traditionally been exclusive to the OGC, most notably the Spatial Data on the Web Working (W3C 2015) and Interest Groups (W3C 2017) and the best practices documents produced in this context (W3C 2019). The formation of these groups leverages the opportunity to involve a much broader group of users in the discussion around how GI should be shared on the web, and their discussions and outputs clearly show that the integration of GI from different sources is a semantic issue at its core (Kuhn 2005). This semantic interoperability and its role for the future of GII in the context of Digital Earth is discussed in detail in Sect. 5.4.2. The following section describes how GII is a prime example of the openness that has become the new normal in many fields of science, technology, and business.

5.3.2 Geospatial Information Infrastructures Champion Openness

As presented earlier, today's GII landscape is shaped by the influence, development and widespread adoption of open data (see also above). The notion of 'openness' has long been part of GIIs, especially due to the long-term leading role of the public sector (Schade et al. 2015). A foundational role of GIIs has been, and still is, to enable the discovery and sharing of spatially referenced data. As described in Sect. 5.2, SDIs were essentially designed and developed to support the generation, management and processing of GI, as key vehicles to make data openly accessible to a broader community. However, as a social construction, the understanding and interpretation of openness is far from static; it is dynamic and changes as the tandem technology-society evolves. Thus, the interpretation of 'open' (data, tools, etc.) reflects the changes in society and necessarily adapts to the new uses and needs of people. In addition, the value of open data is under scrutiny (Craglia and Shanley 2015) and an increasing number of commercial companies produce and host GI.

To better understand how the current discussion about openness affects GIIs and to better speculate future scenarios, we provide two brief stories, paraphrasing the way Arribas-Bel and Reades (2018) examine the evolution between geography and computers. First, we take a brief historical perspective to determine what openness meant in the origins of (governmental) GIIs. Next, we look at the new 'open' trends and growing forces that are currently emerging, mostly outside of GIIs, which we argue are important for GIIs (and Digital Earth) to pay close attention to. Both stories allow for us to reflect and speculate on the need for a convergent point in the future, where GIIs can embrace and continuously adapt to evolving notions of openness and to the resulting societal changes and economic implications.

The first story goes back to the reasons that motivated the need to establish GIIs. Since the outset, GIIs in the form of hierarchical visions on SDI (Rajabifard et al. 2003) or networked visions (Tulloch and Harvey 2008; Vandenbroucke et al. 2009) contained relatively restricted themes and types of resources owned by the public sector. The underlying motto was "collected once, shared multiple times", so each GII node was managed homogeneously its own spatially referenced data. Data sharing was feasible through these infrastructure nodes because data discovery, access, and delivery were affordable through well-known standardization practices (see Sects. 5.2.2 and 5.3.1). Standardized data models and service interfaces characterized the data sharing capabilities of these government-led GIIs, although only a small group of specialized, tech-savvy users benefitted from them. At that time, the concept of openness was tightly coupled to the idea of sharing. The democratization of data sharing through GIIs was a great leap to facilitate transnational and multidisciplinary projects because the problems of discovery, access and redundancy of GI were significantly alleviated by standardized and unified mechanisms. Most recently, this led to the offerings of location enabled e-Services using web-based application programming interfaces (APIs) built upon SDIs. One example of this is the development of an application for citizens called Spotbooking to apply for, process and maintain uses of public spaces within a town or city (Spotbooking 2019).

In addition to past studies to find synergies bridging geospatial research data with public sector information and open data initiatives, other relevant open trends/movements enable knowledge/data collection, creation and dissemination and mostly operate outside GIIs (Schade et al. 2015). We do not list the multitude of open trends and their technological infrastructure here but highlight a few examples to underline the evolving meaning of openness from data sharing to dynamic processes for knowledge production and dissemination. One example is the European Open Science Cloud (EOSC 2019), a cloud for research data in Europe that supports the ongoing transitions in how research is performed and how knowledge is shared. As a second example, the IoT infrastructure generates a vast amount of spatiotemporal data streams at a finer granularity, which undoubtedly represent valuable sources of data (i.e., 'things' observe the environment by collecting data) and analytical computation (i.e., 'things' act by processing gathered data) for Digital Earth and GIIs. Granell et al. feature the promising bridges and synergies between the IoT and Digital Earth application scenarios in Chap. 11, but a true convergence of the two infrastructures is still in its infancy. As a third example, the relationship between Digital Earth and citizen science is outlined in detail in this book by Brovelli and others (Chap. 18).

Fast-forwarding to the present, openness has become a more prominent concept than ever. It has been transformed and extended to all aspects of people's daily lives (Price 2013). Contrary to the common perception of openness in the first example, which was practically restricted to 'sharing' data, today's vision of openness takes multiple and varied forms (Sui 2014). Openness permeates many facets of today's culture, society, government, science and education, leading to a series of (old and new) 'open terms' such as open culture, open cities (Domingo et al. 2013; Degbelo et al. 2016), open movement (Lee et al. 2015), open government (Lathrop and Ruma 2010; Goldsmith and Kleiman 2017), open software (Aksulu and Wade 2010), open hardware (Powell 2012), open science, open research, open laboratories (Nosek et al. 2015), open innovation (Schade and Granell 2014; Mathieu and Aubrecht 2018), and open education (Bonk et al. 2015). In contrast, as analyzed below, daily (geospatial) information still flows to platforms that are not defined as open and are owned by the above-mentioned companies. Offering services free of charge but in exchange for personal (user-generated) data has become a popular business model.

We argue that peoples' perception of openness is dramatically influenced by the irruption, rapid adoption, and new uses and appropriations of technology. Digital transformations brought changes in the proliferation of new data sources, the consolidation of novel ways of producing and consuming data, and in the demography of users. The cost of creating GI anywhere, at any time, from anyone, about anything (aka 4-A technology) drastically decreased. However, the cost for current GIIs to consume, integrate and make sense of 4A-generated data is still considerable—especially when considering the direct and indirect costs for the provision and application of 4-A-generated data for a rich portfolio of use cases and stakeholders (Johnson et al. 2017). The scale, frequency, and granularity of the data being generated and gathered today were simply unimaginable when the foundations of GIIs were designed many years ago. The motto "collected once, shared multiple times" is no longer a fundamental truth that drives GIIs because anyone can collect data on the same phenomenon, in the same place, from multiple perspectives, which was previously technically infeasible. In fact, we unconsciously create such GIs all the time. As a result, more and more data sources are available for a single phenomenon, requiring additional analytical approaches and interoperability arrangements to integrate these data sources and offer a comprehensive picture about the phenomenon in question (Huang et al. 2018). Thus, data in traditional (governmental) GIIs provide one perspective of a phenomenon (mobility, pollution, demography, etc.). Other perspectives of that phenomenon are provided by data that are collected via other infrastructures. This does not fully address the concept of openness. Openness means sharing data about a phenomenon for small groups of experts, enabling and promoting comprehensible views of phenomena taken from disparate sources, and making them accessible and understandable to various user groups. What characteristics do modern GIIs need to fully exploit 4A-generated data? What does this imply in terms of interoperability? And how does this impact current approaches to openness?

While common sense tells us that the way to solve the growing complexity of today's social challenges and underlying research problems is through multidisciplinary collaboration at all levels including technical infrastructures, access to data, and participation in the creation and dissemination of knowledge, the reality is that the diversity of 'open' trends is understandable considering the diversity of actors that have different objectives and needs and are affected differently by a constantly changing technological landscape. It appears that each actor (citizens, NGOs, scientists, private companies, government, etc.) has a different understanding of the meaning and application of the notion of openness. All of them are entirely legitimate given the contexts in which each of the different stakeholders operate.

Regardless of any controversy about the future meaning of openness, it is clear that 'open' cannot be considered a static feature of data or of GII, but should be considered under the lens of recent trends and critiques as a dynamic process for the production, creation and dissemination of knowledge, which is subject to improvements and optimizations over time. The reconceptualization of openness as a dynamic process is vital to enable convergent points and bridges among emerging movements and GIIs—which still operate rather disconnectedly—to make sense of the vast amounts of collected data to solve the pressing issues facing the Earth today. We can rephrase the previous questions: What characteristics would define such dynamic processes in GIIs to exploit 4A-generated data?

Leading GIS scientists recently reflected on the current limitations of the field and called for an entirely new brand of geospatial algorithms and techniques to analyze and process these new forms of data (Jiang 2015; Miller and Goodchild 2015; Li et al. 2016). Lü et al. (2019) magnificently summarize this perception in one sentence: "a successful past [of GII] does not guarantee a bright future" (pp 347). The historical view of GII reported in this chapter is indisputably a story of success. Nevertheless, new driving forces and trends such as open movements and open information infrastructures—along with the datafication and platformization of society—have had and will have significant impacts on the future success of GIIs, so GIIs should carefully consider them to explore alliances and actively integrate and process new forms of information sources.

5.3.3 Capacity Building and Learning for Geospatial Information Infrastructures

Although appropriate technologies and policies to enable data access and data sharing are crucial in the development of GIIs, it also requires education and capacity building to ensure the necessary knowledge, skills and competencies are available

(Craglia et al. 2008). Complementing more general frameworks on the development of digital skills (van Deursen and van Dijk 2014), the need for collaboration between government, businesses and academics in the development of an appropriate knowledge infrastructure has been reflected in national and regional GII strategies and actions (Vancauwenberghe and Vandenbroucke 2016). In the past 20 years, various education and training initiatives on GII and related topics have been developed and implemented by higher education institutions, public administrations and businesses. Throughout the years, the focus has broadly shifted from raising awareness of the potential of GI, to capacity building for the implementation of different GII components to skills and knowledge related to the use and integration of GII data and services in decision making, service delivery and product development processes. GII education and training also must be dynamic and change in response to new technological and policy-related developments. The key challenge in successful GII education and training is to ensure that it addresses the needs of GII professional developers and users. Demand-driven GII education and training requires insight in and agreement on what professionals in the domain of GII should know and be able to do (Vandenbroucke and Vancauwenberghe 2016). Studies investigating the demand for GII capacity building have been undertaken at organizational, national and cross-national levels. A European-wide study on the workforce demand in the domain of GISandT showed that, despite differences in the tasks they perform, employees and representatives from the different sectors including public administration, private sector and academia have strongly similar views on the skills and knowledge areas they consider the most relevant (Wallentin et al. 2014). The European GI community identified a shift in focus from map making and local database handling towards online and mobile technologies based on SDIs with a massive amount of—open—data to be integrated. This is a clear indication that the importance of capacity building for GII will increase in the near future.

A valuable approach in the identification of the specific knowledge and skills that professionals need to master for career success in their field is the development of a comprehensive inventory of the knowledge domain. To provide such an inventory for the GISandT domain, in 2006 the University Consortium of Geographic Information Science (UCGIS) developed the Geographic Information Science and Technology Body of Knowledge (GISandT BoK) (DiBiase et al. 2006). The main intended use of the GISandT BoK was to support the development and assessment of GISandT curricula, but the document also serves other purposes such as for professional accreditation or screening of employees. The 2006 version of the Body of Knowledge included more than 330 topics organized into seventy-three units and ten knowledge areas. Notably, the concept of 'spatial data infrastructure' was included twice, in two different knowledge areas: once in the knowledge area of geospatial data (as a topic under the 'Metadata, standards and infrastructures' unit) and once in the organizational and institutional aspects knowledge area (as a topic under the Institutional and interinstitutional aspects unit). This reflects the need for training and education on the technological and organizational (or institutional) aspects of SDI. In addition to the concept of spatial data infrastructure, the Body of Knowledge contains other concepts that are linked or relevant to the development of SDIs and GIIs, spread across different knowledge areas and units. This demonstrates the relevance and importance of GIIs as a field and the need for an ontology-based approach to the field, where different types of relationships between concepts can be identified (Vandenbroucke and Vancauwenberghe 2016).

To reflect and address recent trends, developments and challenges in the GISandT domain, continuous revision and updating of the Body of Knowledge are required. Initiatives to revise and update the Body of Knowledge have been undertaken and are ongoing in Europe and the United States (Vandenbroucke and Vancauwenberghe 2016). In addition to the topics covered and defined learning objectives, another key aspect in the design and implementation of GII training and education is the teaching and learning activities applied to help students achieve these objectives. GII education has evolved from traditional 'teacher-centric' teaching styles to more 'learner-centric' methods and approaches. With the availability of online—open—education resources by organizations and institutions such as the EuroSDR (EduServ program), the University of Salzburg (UNIGIS program), the Geographical Information System International Group (GISIG) and recently the European Commission (Geospatial Knowledge Base (GKB) Training Platform), the GI/GII community has a strong tradition of e-learning activities. Collaboration between higher education institutions and other stakeholders to design and deliver GI and GII education has taken place for many years. In many cases, this collaboration is often organized in a rather traditional manner, through internships at public or private organizations, the provision of data and tools for educational purposes, and the organization of study visits and excursions to private or public organizations in the GISandT domain (Vancauwenberghe and Vandenbroucke 2016). Recently, several universities started experimenting with more case-based approaches in which students and teachers closely collaborate with practitioners on real-life case studies. The concept of academic SDIs for research and for education can be viewed in the context of adopting more innovative teaching and learning methods (Coetzee et al. 2017). Students could actively contribute to the development and implementation of various SDI components and use the infrastructure to share the results of their efforts with other students, teachers and researchers. In addition, GIIs play a role in the cocreation of knowledge and thereby in lifelong learning (Foresman et al. 2014), and through their fundamental contribution to the Digital Earth vision, GIIs can enable living labs, i.e., user-driven approaches to innovation (Schade and Granell 2014).

5.4 Recent Challenges and Potential for Improvement

Given the situation today—as indicated in the introduction to this chapter—we face a series of challenges. These challenges primarily emerge from the pace of technological change, including more frequent technological disruptions than in the past, and the (to some extent heavy-headed) standardization applied to GIIs. We note the challenges caused by what we call the 'big data' phenomenon (Tsinaraki and Schade 2016) and by the mainstreaming of GI, which introduced new users with new needs as well as new providers of GII. Given these current changes and challenges, we emphasize two implications for GIIs and their future evolution.

5.4.1 Strengthened Role of Semantics

The insight that semantic heterogeneity is a key factor that interferes with the effective use and analysis of GI from different sources is by no means new, nor are the solutions based on semantic web technologies to address the corresponding challenges (Kuhn 2005; Lutz and Klien 2007; Lutz et al. 2009). However, although academic research noted these issues quite early on in the establishment of SDIs, in practice, most efforts have been focused on achieving the underlying technical and syntactic interoperability. This focus is understandable, as semantic interoperability only becomes an issue when the technical and syntactical issues are largely solved. This stage in the development of GII appears to have been reached, since the role of geospatial semantics has been strengthened considerably and is now an issue that practitioners deal with in implementation of open data platforms and geospatial web services.

Arguably, this development was not solely driven by questions about the semantics of geospatial data at hand. Rather, the need for approaches that let us add information about the semantics of entities (geospatial features, in our case), particularly their types and properties, has been recognized in many other fields. These include generic examples such as the publication of structured data on the web (Schema.org is the most prominent example) or specialized application domains (such as biology or history), and closely related research fields such as the sensor web and the Internet of Things. The common need for structured data with clearly defined semantics across those domains has led to efforts in a number of different directions, including research on the theoretical underpinnings of semantic reasoning (Noy 2004; Wang et al. 2004), development of specifications [RDF(S) (Staab et al. 2002), OWL (McGuinness and Van Harmelen 2004), OWL2 (Hitzler et al. 2009; Motik et al. 2009), query languages [SPARQL (Harris et al. 2013), GeoSPARQL (Battle and Kolas 2012)], implementation of the triple stores (Rohloff et al. 2007) and query engines (Broekstra et al. 2002; Carroll et al. 2004). In combination, these efforts have led to a more widespread adoption of approaches that focus on the semantics of geospatial data (Stock et al. 2011), and semantics is now front and center in best practice recommendations for publishing spatial data (W3C 2019).

The W3C's Spatial Data on the Web Best Practices discusses how to best semantically annotate geospatial data—, i.e., using shared vocabularies—and recommends full-fledged adoption of Linked Data principles. A more widespread adoption of these best practices will imply a paradigm shift (Kuhn et al. 2014) towards a radically distributed approach to the publication of GI. Linked Data are currently treated as a byproduct in the publication of GI, e.g., when government agencies such as the UK's Ordinance Survey are starting to offer their GI as Linked Data or when universities convert OSM data to Linked Data. These are valuable efforts—a little semantics goes a long way (Hendler 2009)—but the data that is being published is still the output of an extract-transform-load (ETL) process on top of an original data source such as a relational spatial database. Furthermore, the provision of GI as Linked Data only adds another data offering with the potential use for data integration. Actual success cases remain rare.

The opportunities and challenges of making Linked Data the original data format based on which all changes are made and from which other formats can be derived can currently be observed in the Wikidata effort (Vrandečić and Krötzsch 2014). After the immense success of DBpedia (Auer et al. 2007)—a Linked Data product generated via ETL from the structured information in Wikipedia—the potential of turning this process around by making the produced structured data the actual data underlying all language editions of Wikipedia has been recognized. This approach now allows for an editor to update information in Wikidata—such as the population number for a country after a new census, or the publication of the latest book by a given author—and that information can automatically be reused across all Wikipedia.

GI still has a way to go in making a semantics-based approach its primary format for data management and publication, and thus become part of an ever-growing distributed knowledge graph. Conceived as part of the infrastructure driving Digital Earth, this goal appears attractive, particularly because of its potential to further normalize the use GI across a wider range of disciplines. However, a number of challenges must be addressed before this vision can be put into practice, including the development and implementation of standardized handling of GI in triple stores, interfaces to access geospatial Linked Data directly from GI 'front ends' such as traditional GIS, web-based and mobile mapping applications, as well as capacity building, particularly in the form of educating students in the underlying technology stack so that they can help with these developments after graduation. These challenges highlight the fact that geospatial semantics will remain an essential and dedicated research area for the foreseeable future, helping users make sensible use of GI and turn it into actionable knowledge. Finally, in the context of Linked Data, (geo)spatial information is definitely special because spatial (and particularly spatiotemporal) data can be used as an integrator to help build connections between originally disparate data sources.

5.4.2 Is Spatial Still Special?

There are several slogans related to GI, including "spatial is special". Although one might argue that GI is only more complex than many kinds of (nonspatial) information, at least in the past, geospatial informatics filled a niche role with comparably few specialists working on the topic. As far as mainstream computing was concerned, the spatial-temporal components of GI were restricted to a pair of coordinates (a point) and a date-time stamp. Today, the spatiotemporal characteristics of GI have made it popular for data integration tasks, where location is an obvious commonality between many separately collected data sources (Tsinaraki and Schade 2016). In combination with the recent trend towards platformization of society and wider use of remotely sensed images, online maps, sensors (see also Chap. 11), as well as people's location and tracks, one might argue that the time when (geo)spatial has been special has come to an end. However, although the collection of GI has become much easier (and hence gone mainstream), pitfalls in analysis (spatial autocorrelation, projections, etc.) remain. Related special challenges surface, especially when standard approaches for handling big data are directly applied to GI. Many of the common "divide and conquer" approaches applied to big data analysis tasks fail because of the spatial relationships between chunks of data. As argued in Sect. 5.4.1, semantics is highly important. Using colocation as the only element for data integration can easily lead to the senseless combined processing of data from completely different and potentially conflicting contexts.

The mainstreaming of location information has direct implications for the evolution of GII, as with the future conceptualization of GIS and SDIs. In the past, these notions were a research and application field in their own right, and they now appear to be much more integrated into the wider fields of computer science and data science (Cadell 2018). With a narrow view, this could be seen as a thread to the communities and associations that formed around these concepts (the introduction to this chapter provided some examples of these). Conversely, the mainstreaming of GI provides immense opportunities such as the increasing market for companies specializing in GI and many new job opportunities for GI experts.

From a government perspective, GIIs became more relevant—and geospatial data less special—through the use of data in this infrastructure for the provision of spatially enabled e-government services to citizens, businesses and other societal actors (Vancauwenberghe and van Loenen 2018). Geospatial data that became increasingly available were used to improve existing e-services and provide novel services. Such spatially enabled e-services now exist in many policy areas (i.e., environment, agriculture, transport) and at different levels of government (i.e., local, regional, national). They evolved from more simple information and contact services to more advanced transaction services. These spatially enabled transaction services refer to the use of geospatial data in the electronic intake and handling of requests and applications of rights, benefits and obligations. Because these transaction services demand multiple two-way interactions between governments and citizens/businesses, they are more complex than information or contact services, which are mostly one-way services. This increased complexity applies to both technological and organizational aspects, since the delivery of these e-services requires a strong alignment and possible integration of GIIs with e-government developments. Initiatives to enable this integration have been taken at organizational, national and regional levels—especially in Europe (Vancauwenberghe and van Loenen 2018).

In the private sector, we have observed manifold developments. First, the traditional partnerships with the public sector evolved into collaborations in which governmental bodies such as mapping agencies still own and provide authoritative content (such as cadaster information, protected sites, and utilities), and the industry offers solutions for data hosting, access, and cost recovery. The data and information access services (DIAS) for the European Space program Copernicus is a particularly impressive example (Copernicus 2019). In each of these five different implementations, the public-sector GII is coupled with data from commercial satellites to provide additional value. Second, there are an increasing number of companies building upon GIIs. Especially for technologies such as web-based APIs, as in the example of Spotbooking, GI has become more accessible and value-added services and applications have been created. Due to the abovementioned platformization, large internet firms create many GIs and host them in their infrastructures, and they are only occasionally linked to existing public-sector GIIs. GI has clearly moved into the mainstream information infrastructures. Lastly, many GI projects today rely on data provided by companies such as Google, DigitalGlobe, Waze, Here, and Esri. Examples include geospatial data about commercial demographics and personal mobility.

In the context of Digital Earth, these developments are all good news. In every conceptualization of the Digital Earth vision—and in any future evolution thereof—GI and GIIs will remain fundamental building blocks. As increasing related expertise becomes available and the mainstreaming trend of GI continues, GI can provide the capacity that is required for improving Digital Earth applications and enlarging implementations of the Digital Earth vision across the globe. The transition from mainstreamed GI to GIIs that are readily available to developers and implementers of the Digital Earth vision is the logical next step and an area for further research and organizational improvements. The interplay between and the changes in power relationships between society, research, industry and the public sector deserve dedicated attention.

5.5 Conclusion and Outlook

This chapter situated GIIs in the wider context of the Digital Earth vision and introduced GIIs as a major enabling element for Digital Earth implementation. The past and present of GIIs was outlined along with a subjective view of today's major challenges concerning the status of GII development and use, and possible future directions. Notably, this view might be biased towards academia and governments, but we have highlighted emerging developments from the private sector as a disruptive driving force that quickly emerged over the past decade.

This chapter demonstrates that GIIs have come a long way and evolved as a strong underpinning contribution for implementation of the Digital Earth vision. Whereas we witnessed a dispersion of efforts in the early days, we illustrated how GIIs evolved and coordinated efforts emerged in different national and international contexts. The increasing pace of technological changes poses new challenges to the continuation and further convergence of these efforts because new actors with different backgrounds and expectations enter the discussion. We see a particular need to continue and strengthen the role of semantics in GII development and implementation to ensure that the provided information can be used appropriately. We also recognize the changing power relationship from the public to the private sector, with a disrupting effect on traditional data owners (especially mapping agencies). These changes will significantly affect the role of the public sector in geospatial data management and provision.

Lastly, we underlined the needs for further evolution of GIIs so that they become flexible and robust enough to absorb and embrace technological transformations and the accompanying societal and organizational implications. These required capacities for addressing technological and organizational issues, and training of present and future generations of GII developers and GII users. As a prominent example, we highlighted the relationships to movements to open up data and the access to knowledge. GIIs—which were in the forefront of open data sharing in the past—must react to changing conditions, provide bridges to other existing infrastructures to absorb new data sources, and contribute to the development of new standards for collaboration. The next generation of GIIs should provide management and processing capacities for classical GI, and must be able to input and handle novel information sources. In this way, they will continue to fuel innovation for the future of Digital Earth. Chapters 6, 9 and 10 provide additional insight into analytical aspects and issues related to big data. For details about the economic value of Digital Earth, we refer the reader to Chap. 19.

References

Aksulu A, Wade MR (2010) A comprehensive review and synthesis of open source research. J Assoc Inf Syst 11(11):576–656

ANZLIC (1992) Land information management in Australasia 1990–1992. Australia Government Publishing Service, Canberra

ANZLIC (2019) ANZLIC home page. ANZLIC. https://anzlic.gov.au/. Accessed 6 May 2019

Arribas-Bel D, Reades J (2018) Geography and computers: past, present, and future. Geogr Compass 12(10):e12403

Auer S, Bizer C, Kobilarov G et al (2007) DBpedia: a nucleus for a web of open data. In: Aberer K, Choi K-S, Noy N et al (eds) The semantic web. ISWC 2007, ASWC 2007. Lecture notes in computer science, vol 4825. Springer, Heidelberg, pp 722–735

Battle R, Kolas D (2012) Enabling the geospatial semantic web with parliament and geoSPARQL. Semant Web 3(4):355–370

Berners-Lee T (2006) Linked data: design issues. W3C. https://www.w3.org/DesignIssues/LinkedData.html. Accessed 6 May 2019

Bonk CJ, Lee MM, Reeves TC et al (2015) MOOCs and open education around the world. Routledge, New York

Bostock M, Ogievetsky V, Heer J (2011) D3: data-driven documents. IEEE Trans. Visualization and Comp. Graphics (Proc. InfoVis). Stanford Vis Group, Stanford

Broekstra J, Kampman A, van Harmelen F (2002) Sesame: a generic architecture for storing and querying RDF and RDF schema. In: Horrocks I, Hendler J (eds) International semantic web conference, the semantic web — ISWC 2002. Springer, Heidelberg, pp 54–68

Brovelli MA, Minghini M, Moreno-Sanchez R et al (2017) Free and open source software for geospatial applications (FOSS4G) to support future earth. Int J Digital Earth 10(4):386–404

Burrough P, Brand M, Salge F et al (1993) The EUROGI vision. GIS Eur 2(3):30–31

Cadell W (2018) Geospatial analytics will eat the world, and you won't even know it. Forbes Technology Council. https://www.forbes.com/sites/forbestechcouncil/2018/07/16/geospatial-analytics-will-eat-the-world-and-you-wont-even-know-it/#40457209726a. Accessed 6 May 2019

Capineri C, Haklay M, Huang H et al (2016) European handbook of crowdsourced geographic information. Ubiquity Press, London

Carroll JJ, Dickinson I, Dollin C et al (2004) Jena: implementing the semantic web recommendations. In: Feldman S, Uretsky M, Najork M et al (eds) Proceedings of the 13th international World Wide Web conference on alternate track papers and posters. ACM, New York, pp 74–83

Castells M, Cardoso G (2005) The network society: from knowledge to policy. Johns Hopkins Center for Transatlantic Relations, Washington DC

Coetzee S, Steiniger S, Köbben B et al (2017) The academic SDI—Towards understanding spatial data infrastructures for research and education. In: Peterson MP (ed) International cartographic conference. ICACI 2017: advances in cartography and GIScience. Springer International Publishing, Cham, pp 99–113

Copernicus (2019) DIAS EC. https://www.copernicus.eu/en/access-data/dias. Accessed 6 May 2019

Coppock JT (1987) Handling geographic information. Report of the committee of enquiry. Int J Geogr Inf Syst 1(2):191–192

Craglia M, Goodchild MF, Annoni A et al (2008) Next-generation digital earth: a position paper from the Vespucci initiative for the advancement of geographic information science. Int J Spat Data Infrastruct Res 3:146–167

Craglia M, Shanley L (2015) Data democracy – increased supply of geospatial information and expanded participatory processes in the production of data. Int J Digit Earth 8(9):679–693

Danko D (2005) ISO/TC 211 metadata. Geo-information standards in action. Nederlandse Commissie voor Geodesie (NCG), Delft

Davis Jr. CA (2009) Spatial data infrastructures. In: Khosrow-Pour M (ed) Encyclopedia of information science and technology, second edition. IGI Global, Hershey, pp 3548–3553

Degbelo A, Granell C, Trilles S et al (2016) Opening up smart cities: citizen-centric challenges and opportunities from GIScience. ISPRS Int J Geo-Inf 5(2):16

DiBiase D, DeMers M, Johnson A et al (2006) Geographic information science and technology body of knowledge. Association of American Geographers, Washington DC

Domingo A, Bellalta B, Palacin M et al (2013) Public open sensor data: revolutionizing smart cities. IEEE Technol Soc Mag 32(4):50–56

Dragoni N, Giallorenzo S, Lafuente AL et al (2017) Microservices: yesterday, today, and tomorrow. In: Mazzara M, Meyer B (eds) Present and ulterior software engineering. Springer International Publishing, Cham, pp 195–216

EOSC (2019) European open science cloud (EOSC). EC. http://ec.europa.eu/research/openscience/index.cfm?pg=open-science-cloud. Accessed 6 May 2019

ESRI (1998) ESRI shapefile technical description: an ESRI white paper. Environmental Systems Research Institute, Redlands

ESRI (2019a) ArcGIS hub examples. In: ArcGIS Hub. ESRI. Available via DIALOG. https://hubdemo-cityx.opendata.arcgis.com/. Accessed 6 May 2019

ESRI (2019b) ArcGIS API for JavaScript. In: JavaScript API, Version 4.12. ESRI. Available via DIALOG. https://developers.arcgis.com/javascript/. Accessed 6 May 2019

ESRI (2019c) ArcGIS open data charter. ESRI. https://www.esri.com/en-us/arcgis/open-vision/standards/open-data. Accessed 15 Apr 2019

Executive Office of the President (1994) Executive Order 12906: Coordinating Geographic Data Acquisition and Access: the National Spatial Data Infrastructure, signed by President Bill Clinton on April 11, 1994, launched the initiative to create the NSDI. https://www.fgdc.gov/policyandplanning/executive_order. Accessed 12 Aug 2019

FGDC (1998) Content standard for digital geospatial metadata, FGDC-STD-001-1998. Federal Geographic Data Committee, Metadata Ad Hoc Working Group, Washington DC

Foresman T, Schade S, Georgiadou Y et al (2014) Does DE need a C? A proposal for a DE curriculum. Int J Digit Earth 7(1):88–92

Geospatial World (2011) Geospatial world forum 2011: technology for people and earth, geospatial world. Geospatial World. https://www.geospatialworld.net/article/geospatial-world-forum-2011-

technology-for-people-and-earth/. Accessed 6 May 2019
Gimpel H, Röglinger M (2015) Digital transformation: changes and chances – Insights based on an empirical study. Fraunhofer FIT, Sankt Augustin
Goldsmith S, Kleiman N (2017) A new city O/S: the power of open, collaborative, and distributed governance. Brookings Institution Press, Washington DC
Goodchild MF (2007) Citizens as voluntary sensors: spatial data infrastructure in the world of web 2.0. Int J Spat Data Infrastruct Res 2:24–32
Goodchild MF, Guo H, Annoni A et al (2012) Next-generation digital earth. Proc Nat Acad Sci USA 109(28):11088–11094
Google (2019) Google Maps home page. Google. https://www.google.com/maps. Accessed 6 May 2019
Gore A (1998) The digital earth: understanding our planet in the 21st century. Speech given at the California Science Center, Los Angeles, California, on January 31ESRI. http;//portal.opengeospatial.org/files/?artifact_id = 6210. Accessed 6 May 2019
Granell C, Fernández ÓB, Díaz L (2014) Geospatial information infrastructures to address spatial needs in health: collaboration, challenges and opportunities. Future Gener Comput Syst 31:213–222
Granell C, Havlik D, Schade S et al (2016) Future internet technologies for environmental applications. Environ Model Softw 78:1–15
GRASS (2019) GRASS GIS home page. OSGeo. https://grass.osgeo.org/. Accessed 6 May 2019
Gurstein MB (2011) Open data: empowering the empowered or effective data use for everyone? First Monday 16(2). https://doi.org/10.5210/fm.v16i2.3316
Haklay M, Weber P (2008) OpenStreetMap: user-generated street maps. IEEE Pervasive Comput 7(4):12–18
Harris S, Seaborne A, Prud'hommeaux E (2013) SPARQL 1.1 query language. W3C Recomm 21(10):806
Hendler J (2009) A little semantics goes a long way. Rensselear School of Science. https://www.cs.rpi.edu/~hendler/LittleSemanticsWeb.html. Accessed 6 May 2019
Hitzler P, Krötzsch M, Parsia B et al (2009) OWL 2 web ontology language primer. W3C Recomm 27(1):123
Huang H, Gartner G, Krisp JM et al (2018) Location based services: ongoing evolution and research agenda. J Locat Based Serv 12(2):63–93
ILWIS (2019) Integrated land and water information system (ILWIS). ITC, University of Twente. https://www.itc.nl/ilwis/. Accessed 6 May 2019
INSPIRE (2019) INSPIRE home page. EC. https://inspire.ec.europa.eu/. Accessed 6 May 2019
ISO (2002) ISO 19101:2002 geographic information – Reference model. ISO. https://www.iso.org/standard/26002.html. Accessed 6 May 2019
ISO (2003) ISO 19115:2003 geographic information – Metadata. ISO. https://www.iso.org/standard/26020.html. Accessed 6 May 2019
Jiang B (2015) Geospatial analysis requires a different way of thinking: the problem of spatial heterogeneity. GeoJournal 80(1):1–13
Johnson PA, Sieber R, Scassa T et al (2017) The cost(s) of geospatial open data. Trans GIS 21(3):434–445
Kitchin R (2014) The data revolution: big data, open data, data infrastructures and their consequences. SAGE Publications Ltd, Maynooth
Kuhn W (2005) Geospatial semantics: why, of what, and how? In: Spaccapietra S, Zimányi E (eds) Journal on data semantics III. Springer, Heidelberg, pp 1–24
Kuhn W, Kauppinen T, Janowicz K (2014) Linked data - A paradigm shift for geographic information science. In: Duckham M, Pebesma E, Stewart K et al (eds) International conference on geographic information science. Springer International Publishing, Cham, pp 173–186
Lathrop D, Ruma L (2010) Open government: collaboration, transparency, and participation in practice. O'Reilly Media, Sebastopol

Leaflet (2019) Leaflet home page. Leaflet. Open Streat Map. http://leafletjs.com. Accessed 6 May 2019
Lee M, Almirall E, Wareham J (2015) Open data and civic apps: first-generation failures, second-generation improvements. Commun ACM 59(1):82–89
Li S, Dragicevic S, Castro FA et al (2016) Geospatial big data handling theory and methods: a review and research challenges. ISPRS J Photogramm Remote Sens 115:119–133
Longley PA, Goodchild MF, Maguire DJ et al (2011) Geographic information systems and science. John Wiley and Sons, Hoboken
Lü G, Batty M, Strobl J et al (2019) Reflections and speculations on the progress in geographic information systems (GIS): a geographic perspective. Int J Geogr Inf Sci 33(2):346–367
Lutz M, Klien E (2007) Ontology-based retrieval of geographic information. Int J Geogr Inf Sci 20(3):233–260
Lutz M, Sprado J, Klien E et al (2009) Overcoming semantic heterogeneity in spatial data infrastructures. Comput Geosci 35(4):739–752
Maguire DJ, Longley PA (2005) The emergence of geoportals and their role in spatial data infrastructures. Comput Environ Urban Syst 29(1):3–14
Masser I (2005) GIS worlds: creating spatial data infrastructures. ESRI Press, Redlands
Mathieu PP, Aubrecht C (2018) Earth observation open science and innovation. Springer Open, Cham
McGuinness DL, Van Harmelen F (2004) OWL web ontology language overview. W3C Recomm 10(10):2004
Miller HJ, Goodchild MF (2015) Data-driven geography. GeoJournal 80(4):449–461
Motik B, Grau BC, Horrocks I et al (2009) OWL 2 web ontology language profiles. W3C Recomm 27:61
National Research Council, Division on Earth and Life Studies, Commission on Geosciences et al (1993) Toward a coordinated spatial data infrastructure for the nation. National Academies Press, Washington DC
Nativi S, Craglia M, Pearlman J (2013) Earth science infrastructures interoperability: the brokering approach. IEEE J Sel Top Appl Earth Obs Remote Sens 6(3):1118–1129
Nebert DD (ed) (2004) Developing spatial data infrastructures: the SDI cookbook. Global Spatial Data Infrastructures Association, Reston
Nosek BA, Alter G, Banks GC et al (2015) Promoting an open research culture. Science 348(6242):1422–1425
Noy NF (2004) Semantic integration: a survey of ontology-based approaches. ACM Sigmod Rec 33(4):65–70
NSDI (2019) Advancement of the national spatial data infrastructure. FGDC. https://www.fgdc.gov/nsdi. Accessed 6 May 2019
OGC (2019) Open geospatial consortium home page. OGC. http://www.opengeospatial.org/. Accessed 6 May 2019
OMB (1990) Coordination of surveying, mapping and related spatial data activities. Executive Office of the President, Circular A-16 Revise, Office of Management and Budget, Washington, DC
Open Data Barometer (2015) Open data barometer – Key findings open data barometer. https://opendatabarometer.org/2ndEdition/summary/. Accessed 6 May 2019
OpenLayers (2019) OpenLayers home page. OpenLayers. http://openlayers.org. Accessed 6 May 2019
OSM (2019a) Open Street Map home page. OSM. https://www.openstreetmap.org. Accessed 6 May 2019
OSM (2019b) Slippy map tilenames. OSM. https://wiki.openstreetmap.org/wiki/Slippy_map_tilenames. Accessed 6 May 2019
OSM (2019c) OpenStreetMap stats reports. OSM. https://www.openstreetmap.org/stats/data_stats.html. Accessed 6 May 2019

Papazoglou MP, van den Heuvel W-J (2007) Service oriented architectures: approaches, technologies and research issues. VLDB J 16(3):389–415
Powell A (2012) Democratizing production through open source knowledge: from open software to open hardware. Media Cult Soc 34(6):691–708
Price D (2013) Open: how we'll work, live and learn in the future. Crux Publishing, Great Britain
Rajabifard A, Feeney M-EF, Williamson IP (2002) Future directions for SDI development. Int J Appl Earth Obs Geoinf 4(1):11–22
Rajabifard A, Feeney MF, Williamson I (2003) Spatial data infrastructures: concept, nature and SDI hierarchy. In: Williamson I, Rajabifard A, Feeney MF (eds) Developing spatial data infrastructures: from concept to reality. CRC Press, Boca Raton, pp 17–40
Rohloff K, Dean M, Emmons I et al (2007) An evaluation of triple-store technologies for large data stores. In: Meersman R, Tari Z, Herrero P (eds) OTM confederated international conferences on the move to meaningful internet systems. Springer, Heidelberg, pp 1105–1114
Schade S, Granell C (2014) Shaping digital earth applications through open innovation – Setting the scene for a digital earth living lab. Int J Digit Earth 7(7):594–612
Schade S, Granell C, Perego A (2015) Coupling public sector information and public-funded research data in Europe: a vision of an open data ecosystem. In: Reddick C, Anthopoulos L (eds) Information and communication technologies in public administration: innovations from developed countries. CRC Press, London, pp 275–298
Schade S, Smits P (2012) Why linked data should not lead to next generation SDI. In: 2012 IEEE international geoscience and remote sensing symposium, IEEE, Munich, 22–27 July 2012
Sjoukema J-W, Bregt A, Crompvoets J (2017) Evolving spatial data infrastructures and the role of adaptive governance. ISPRS Int J Geo-Inf 6(8):254
smeSpire (2014) A European community of SMEs built on environmental digital content and languages. smeSpire. http://www.smespire.eu/wp-content/uploads/downloads/2014/03/D1.3_FinalReport_1.0.pdf. Accessed 6 May 2019
Spotbooking (2019) The smart way to reserve & manage the use of public domain. GEOSPARC. https://www.geosparc.com/en/spotbooking. Accessed 6 May 2019
Staab S, Erdmann M, Maedche A et al (2002) An extensible approach for modeling ontologies in RDF(S). In: Grütter R (ed) Knowledge media in healthcare: opportunities and challenges. IGI Global, Hershey, pp 234–253
Steiniger S, Bocher E (2009) An overview on current free and open source desktop GIS developments. Int J Geogr Inf Sci 23(10):1345–1370
Stock K, Hobona G, Granell C et al (2011) Ontology-based geospatial approaches for semantic awareness in earth observation systems. In: Ashish N, Sheth AP (eds) Geospatial semantics and the semantic web: foundations, algorithms, and applications. Springer US, Boston, pp 97–118
Sui D (2014) Opportunities and impediments for open GIS. Trans GIS 18(1):1–24
Tait MG (2005) Implementing geoportals: applications of distributed GIS. Comput Environ Urban Syst 29(1):33–47
Tsinaraki C, Schade S (2016) Big data—A step change for SDI. Int J Spat Data Infrast Res 11:9–19
Tulloch DL, Harvey F (2008) When data sharing becomes institutionalized: best practices in local government geographic information relationships. URISA J 19:51–59
van Deursen AJAM, van Dijk JAGM (2014) Digital skills: unlocking the information society. Palgrave Macmillan US, New York
van Dijck J, Poell T, de Waal M (2018) The platform society: public values in a connective world. Oxford University Press, New York
van Oosterom P (2005) Geo-information standards in action. Netherlands Geodetic Commission (NCG), Delft
Vancauwenberghe G, van Loenen B (2018) Exploring the emergence of open spatial data infrastructures: analysis of recent developments and trends in Europe. In: Saeed S, Ramayah T, Mahmood Z (eds) User centric E-government: challenges and opportunities. Springer International Publishing, Cham, pp 23–45

Vancauwenberghe G, Vandenbroucke D (2016) Collaborative learning: analysis of the needs of industry and academia. http://www.gicases.eu/download/d1-1-collaborative-learning-analysis-of-the-needs-of-industry-and-academia/. Accessed 6 May 2019
Vandenbroucke D, Crompvoets J, Vancauwenberghe G et al (2009) A network perspective on spatial data infrastructures: application to the sub-national SDI of Flanders (Belgium). Trans GIS 13(s1):105–122
Vandenbroucke D, Vancauwenberghe G (2016) Towards a new body of knowledge for geographic information science and technology. Micro Macro Mezzo Geoinf 2016(6):7–19
Vrandečić D, Krötzsch M (2014) Wikidata: a free collaborative knowledgebase. Commun ACM 57(10):78–85
W3C (2014) Data catalog vocabulary (DCAT) W3C. https://www.w3.org/TR/vocab-dcat/. Accessed 6 May 2019
W3C (2015) Spatial data on the web working group, main page. W3C. https://www.w3.org/2015/spatial/wiki/Main_Page. Accessed 6 May 2019
W3C (2017) Spatial data on the web, interest group page. W3C. https://www.w3.org/2017/sdwig/. Accessed 6 May 2019
W3C (2019) Spatial data on the web best practices. W3C. https://www.w3.org/TR/sdw-bp/. Accessed 6 May 2019
Wallentin G, Hofer B, Traun C (2014) Analysis of the demand for geospatial education and training. Results of the GI-N2K demand survey. Deliverable of the project GI-N2K – Geographic Information Need to Know. http://www.gi-n2k.eu/?wpdmdl=455. Accessed 12 Aug 2019
Wang XH, Zhang DQ, Gu T et al (2004) Ontology based context modeling and reasoning using OWL. In: Proceedings of the second IEEE annual conference on pervasive computing and communications workshops, 2004. IEEE, Orlando, 14–17 March 2004
Worboys MF, Duckham M (2004) GIS: a computing perspective, second edition. CRC Press, Boca Raton
Yang C, Raskin R, Goodchild M et al (2010) Geospatial cyberinfrastructure: past, present and future. Comput Environ Urban Syst 34(4):264–277
Zhao P, Di L (2011) Geospatial web services: advances in information interoperability. IGI Global, Hershey

Sven Schade is a scientific officer at the European Commission's scientific and knowledge management service, the Joint Research Centre (JRC), with a background in geospatial information science and knowledge engineering. His interests cover knowledge extraction from heterogeneous data sources, multi-disciplinary interoperability, and public participation in science for policy.

Carlos Granell currently holds a 5 year Ramón y Cajal post-doctoral fellowship at the Universitat Jaume I of Castellón, Spain. His interests lie in multi-disciplinary application of GIS (Science/Systems), spatial analysis & visualization of streams of sensor- and/or user-generated geographic content, reproducibility research practices, and finding synergies between citizen science projects, education and dissemination of science.

Glenn Vancauwenberghe is a senior researcher at the Spatial Applications Division Leuven (SADL) of KU Leuven, with a background in sociology and public administration. His main fields of research are the governance and performance of spatial data infrastructures and the role of geospatial data in open and digital government.

Carsten Keßler is an Associate Professor for Geoinformatics in the Department of Planning at Aalborg University Copenhagen, Denmark, and an adjunct faculty member at The Graduate Center, City University of New York, USA. He is interested in data interoperability, volunteered geographic information, and geo-simulation techniques for large-scale assessments of the impacts of global warming.

Danny Vandenbroucke is research manager at KU Leuven, Belgium. His research interest relates to organizational and technical interoperability aspects and the impact geospatial standards might have on the performance of business processes. He works on several Erasmus+ projects that address the educational challenges in the geospatial domain (defining an ontology-based Body of Knowledge for GI/EO, case-based learning …).

Ian Masser has held professorial positions at the Universities of Utrecht, Sheffield and Twente as well as visiting positions in University College London, Melbourne and Leuven. His publications include twenty books and more than 300 papers. In 2016 he received the Global Citizen Award from the GSDI Association for his exemplary service to the geospatial community.

Michael Gould is a professor of GIS at Universitat Jaume I of Castellón, Spain, and ESRI's Global Education Manager working mostly in Europe, Middle East and Africa. His research interests include distributed GIS, Smart Cities, and hybrid FOSS/proprietary interoperability.

Permissions

All chapters in this book were first published by Springer; hereby published with permission under the Creative Commons Attribution License or equivalent. Every chapter published in this book has been scrutinized by our experts. Their significance has been extensively debated. The topics covered herein carry significant information for a comprehensive understanding. They may even be implemented as practical applications or may be referred to as a beginning point for further studies.

The contributors of this book come from diverse backgrounds, making this book a truly international effort. We would like to thank all the contributing authors for lending their expertise to make the book truly unique. They have played a crucial role in the development of this book. Without their invaluable contributions this book wouldn't have been possible. They have made vital efforts to compile up to date information on the varied aspects of this subject to make this book a valuable addition to the collection of many professionals and students.

This book was conceptualized with the vision of imparting up-to-date and integrated information in this field. To ensure the same, a matchless editorial board was set up. Every individual on the board went through rigorous rounds of assessment to prove their worth. After which they invested a large part of their time researching and compiling the most relevant data for our readers.

The editorial board has been involved in producing this book since its inception. They have spent rigorous hours researching and exploring the diverse topics which have resulted in the successful publishing of this book. They have passed on their knowledge of decades through this book. To expedite this challenging task, the publisher supported the team at every step. A small team of assistant editors was also appointed to further simplify the editing procedure and attain best results for the readers.

Apart from the editorial board, the designing team has also invested a significant amount of their time in understanding the subject and creating the most relevant covers. They scrutinized every image to scout for the most suitable representation of the subject and create an appropriate cover for the book.

The publishing team has been an ardent support to the editorial, designing and production team. Their endless efforts to recruit the best for this project, has resulted in the accomplishment of this book. They are a veteran in the field of academics and their pool of knowledge is as vast as their experience in printing. Their expertise and guidance has proved useful at every step. Their uncompromising quality standards have made this book an exceptional effort. Their encouragement from time to time has been an inspiration for everyone.

The publisher and the editorial board hope that this book will prove to be a valuable piece of knowledge for students, practitioners and scholars across the globe.

Index

A

Abstract Specification, 46, 48-49, 51

B

Basic Framework, 1-3

Broadcast Ephemeris, 130-131, 134-135, 143, 147, 151

C

Capacity Building, 174-175, 178

Cartography, 2, 4, 21, 28, 50, 68, 83, 182

Cell Type, 26

Climate Change, 3-4, 15-16, 18, 53, 75, 110-111, 117, 119, 167

Constellation, 2, 62, 66, 71, 73, 77, 79-80, 120, 125-131, 133, 135-136

Control Segment, 126, 130-131, 134-135

D

Deployment, 126-127, 135

Digital Earth, 1-7, 9-13, 15-21, 24-25, 32-33, 36, 40, 43-46, 50-54, 118, 120-124, 147, 158-160, 167-169, 171-173, 176, 178, 180-183, 185

Digital Earth Initiative, 2, 5-6

Disaster Reduction, 18, 71, 73

Disaster Warning, 75, 113

Discrete Global Grid, 10, 21, 24-25, 38, 46, 50-52

E

Earth Observation, 1-4, 6-8, 11-12, 14-15, 17, 19-20, 46, 54-57, 64, 70, 73-74, 76-78, 80-81, 85-86, 89-90, 115, 118-122, 124, 153, 163-164, 184-185

Equator, 58, 60, 89, 106, 111-112, 114, 134

F

Field of View, 73, 107, 111

Finer Cells, 27

Flood, 7, 18, 112

G

Geographic Information Systems, 44, 50, 158-159, 161, 164, 184

Geography, 2, 4, 20, 98, 172, 181, 184

Geospatial Information, 3, 6, 9, 13-14, 17, 21, 49, 158-159, 161, 164-165, 169, 171, 174, 182-183, 186

Geospatial World, 163-164, 167, 182

Geostationary Orbit, 110, 113, 117

Global Challenge, 2

Global Response, 1-2, 5, 16

Ground Pixel Resolution, 73, 76

H

Hexagonal Cells, 26, 28, 43

Hexagonal Refinement, 27, 37-38

High-precision Positioning, 123, 146-147, 151, 153

I

Imaging Satellite, 56, 62-63, 75-76

Indexing Mechanisms, 33, 38

Indexing Method, 33-36, 50

Indexing Scheme, 38-40, 43

Indexing System, 24, 33, 35-38, 42, 44, 51

Information Infrastructures, 13, 158, 161, 169, 171, 174, 180, 183

Infrared Channels, 107-109, 118

Ionospheric Delay, 143, 152-153

L

Longitude Grids, 26, 40, 42

M

Meteorological Satellite, 85, 89, 99, 102, 104, 106-107, 109-114, 116-117

Meteorological Services, 111-112

Meteorology, 2, 115

Microwave Radiometer, 87, 94, 96-97

Multisource Data, 11, 63

Multispectral Cameras, 82

N

Natural Disaster, 58, 117

O

Ocean Color Satellite, 85

Oceanography, 10, 51, 68, 98

Octahedron, 26, 32, 34, 36-38, 52

Onboard Instrument, 114

Optical Sensor, 80

P

Platformization, 169, 174, 179-180

Polarization Mode, 74, 98

Polyhedron, 25-27, 31, 33, 35-36, 47

Positioning System, 123-124, 126, 137, 141-143, 145, 155, 157

Public Sector, 158-159, 165-167, 171-172, 179-181, 185

Q

Quantization, 26, 48, 69, 94, 101, 107, 109, 114

R

Radio Positioning, 87, 147

Radiometer, 60, 80, 87, 89, 94-97, 103, 106-114, 116-118

Reference Network, 145, 152

Reference Stations, 137-139, 141, 144-146, 152-153

Refinement, 25-28, 33-38, 41, 43-44, 47

Remote Sensing, 1, 4, 6, 8-9, 21, 24, 44, 52-58, 62, 65-66, 70-71, 74-79, 81, 85-87, 95, 98, 109, 113, 118-119, 121-122, 185

Remote Sensor, 60, 63, 83, 95

Resource Investigation, 58, 71, 156

S

Satellite Platform, 89, 110

Satellite Systems, 86, 114-115, 123-125, 157

Scan Line, 60, 103, 111

Space Environment Monitor, 70, 110, 112-113

Spatial Data, 6, 8, 11, 13-14, 21, 44, 48-52, 158-164, 166-168, 171, 175, 177, 182-186

Spatial Resolution, 47, 54, 56-57, 62, 69, 71, 73-76, 79, 82-83, 87, 93, 96, 108, 114, 118-119

Spectral Resolution, 57, 66, 70, 83

Standardization, 17, 46, 158, 160, 165-166, 169-172, 176

Sustainable Development, 2-3, 9, 15-19, 21, 65

T

Technical Parameters, 67, 70, 88, 106-109, 111, 113

Time System, 136, 148

Tropospheric Delay, 152-153

Typhoons, 109, 111-112

U

User Segment, 126, 131

V

Veripos, 138, 142, 144

Visible Light, 54-55, 94-95, 98, 103, 106-107, 109, 111, 113-114, 116-118

Visualization, 2, 5, 9-12, 42, 44-45, 50, 52-53, 160, 171, 181, 186

W

Water Resource, 75, 119

Weather Prediction, 87, 110, 113, 146

Printed in the USA
CPSIA information can be obtained
at www.ICGtesting.com
JSHW010843251023
50683JS00016B/20

9 781641 167796